SOCIETY 3.0

WITHDRAWN

Barbara Bush   Library Friends

Building a better library   *one book at a time.*

In Honor Of

Rhoda Goldberg

Director, Harris County Public Library

This book is part of both the Peter Lang Education list
and the Media and Communication list.
Every volume is peer reviewed and meets
the highest quality standards for content and production.

PETER LANG
New York • Washington, D.C./Baltimore • Bern
Frankfurt • Berlin • Brussels • Vienna • Oxford

TRACEY WILEN-DAUGENTI

SOCIETY 3.0

HOW TECHNOLOGY IS RESHAPING EDUCATION, WORK AND SOCIETY

PETER LANG
New York • Washington, D.C./Baltimore • Bern
Frankfurt • Berlin • Brussels • Vienna • Oxford

Library of Congress Cataloging-in-Publication Data
Wilen-Daugenti, Tracey.
Society 3.0: how technology is reshaping education,
work and society / Tracey Wilen-Daugenti.
p. cm.
Includes bibliographical references and index.
1. Technological innovations—Social aspects.
2. Technological innovations—Economic aspects.
3. Education—Effect of technological innovations on.
4. Employees—Effect of technological innovations on.
5. Technology—Social aspects.
6. Technology and civilization. I. Title.
HM846.W55   303.48'3—dc23   2011032864
ISBN 978-1-4331-1692-6 (hardcover)
ISBN 978-1-4331-1691-9 (paperback)
ISBN 978-1-4539-0219-6 (e-book)

Bibliographic information published by **Die Deutsche Nationalbibliothek**.
Die Deutsche Nationalbibliothek lists this publication in the "Deutsche
Nationalbibliografie"; detailed bibliographic data is available
on the Internet at http://dnb.d-nb.de/.

CONTENTS

Section III—Technology Trends

Section IV—Higher Education and Implications

ACKNOWLEDGMENTS

This book emerged from a passion for innovation and a vision of its power to shape every aspect of our lives. Many people shared my passion and supported my vision. My sincere thanks to Caroline Molina-Ray, PhD, for her foresight into what this book could become and to her publications team for their constant support in helping me articulate and refine the book's core messages. My deep gratitude goes to Sangeet Duchane for her expertise and interest in historic events that provide visibility into future trends; to Courtney L. Vien, PhD, for authoring chapter 2 on working learners and providing an important perspective on the students of today and tomorrow; to Sunanda Vittal for her careful editing and for helping each chapter reflect a deeper understanding of current and future societal issues; to James M. Fraleigh for precise copyediting and smooth integration of the research components; to Graham B. Smith for his creative cover design and graphics; and to Sheila Bodell for her expertise in the areas of library science, academic book indexing, and publishing operations. Their contributions as a virtual team provide a current-day preview into the future of work as a creative, productive, and rewarding endeavor.

SECTION I

SOCIETAL TRENDS

This section discusses key societal changes occurring in the family unit and how a new set of learners is participating in education and the workforce. Traditional family models are giving way to blended family structures in which caregivers include not just a mother and father but also single parents, same-gender couples, siblings, grandparents, and the like. Balancing work and family is a priority for most Americans, and they look to their workplace environment to recognize this need.

Changing family structures have impacted the way Americans perceive higher education goals. Today, there are more nontraditional learners than ever before. They include first-time college entrants with part-time or full-time jobs, sole wage-earners with families, and adult learners extending their careers or—as is increasingly the case—taking on new careers well past traditional retirement age. Changing societal attitudes regarding work and family impact the reasons why people choose to pursue higher education programs. Higher education institutions must therefore factor these changes into the way knowledge is delivered and consumed.

· 1 ·

RUNAWAY CHANGES

America's "traditional" family unit for much of the 20th century—a husband as sole wage-earner, a stay-at-home mother, and two or three children—has radically evolved into multiple new forms. A lively nationwide patchwork of family arrangements and perspectives now thrives alongside the older model. Today's typical families include single-parent households, blended families with step- or half-siblings and stepparents, couples with adoptive or foster children (or children from a surrogate mother), mixed-race households, multigeneration groups with middle-aged caregivers tending elderly and younger members, unmarried or gay couples with or without children, and, of course, dual-parent households with biological children.

These new models have arisen over the past few decades in response to shifting cultural norms and urgent economic realities. In turn, they exert new influence and pressure upon society as their numbers rise and their needs become more pronounced. As the worlds of business, technology, and education respond to the way the members of these new units now work, communicate, and learn, society will come to reflect this diversity and support the American family in all of its many forms.

The Changing Family Landscape

There was a time when the sitcom picture of the American family having just one wage-earner—the man—was fairly accurate. In 1938, up to 68% of U.S. households featured this arrangement (Chinhui & Potter, 2006). Women briefly entered traditional men's jobs during World War II, but once the men returned, women often gave the jobs back to the returning men and began to start families. Those women who chose a career were the exception, and as a result, Rosie the Riveter, an archetype of World War II women factory workers, gave way to perfect housewives like June Cleaver of TV sitcom fame.

In the 1970s a new ideology emerged, promoting the idea that a woman's place was not limited to the home and that a man did not bear the full responsibility of supporting the family. These changing attitudes profoundly affected the American workplace. By 1975, the number of households with two wage-earners (dual-earner households) began to increase steadily until it reached 38% of all U.S. households in 2006. By then, only 16% of U.S. households still had a sole male wage-earner (Chinhui & Potter, 2006), and rising inflation during tougher economic times impelled many families to become dual-earner households just to pay the bills (American Psychological Association, 2004).

Stay-at-Home Parents Today

The United States had an estimated 5.3 million stay-at-home parents in 2009: 5.1 million mothers and 158,000 fathers. The number of stay-at-home moms was lower in 2009 (5.1 million) than in 2008 (5.3 million). The number of stay-at-home dads did not differ statistically between 2008 and 2009.

[U.S. Census Bureau. (2010, January 15). Census Bureau reports families with children increasingly face unemployment [Press release]. Retrieved from http://www.census.gov/ newsroom/releases/archives/families_households/cb10–08.html]

A significant outcome of the rise of the dual-income household has been that couples and families have had to craft new ways to schedule activities, resulting in new lifestyles and responsibilities. Everyday chores like cooking, cleaning, and caring for kids, for instance, must now be worked into job schedules, commuting, and travel. The challenges of handling these mixed roles and responsibilities remain an ongoing negotiation. Maria Shriver, broadcast journalist and author of a report on women's role in society (Shriver & The Center for American Progress, 2009), states that when she interviewed couples about this issue in 2009, many said that they sit down to discuss responsibilities and chores several times each week to keep their lives organized and their relationships viable.

Dramatic Shifts in Demographics and Attitudes toward Marriage

Lifestyle changes among American families have altered the traditional marriage equation. In 1950, there were 2.6 divorces per 1,000 people. In 2001, the rate had almost doubled to 4 per 1,000 (Fields, 2004). In 2006, a significant jump occurred; it was reported that 10.6%, or approximately one tenth of the U.S. population over the age of 15, was divorced (U.S. Census Bureau, 2008).

Statistics on marriage in American households vary slightly, but all show the same trend toward fewer married householders. According to the U.S. Census Bureau (2008), 71% of American households were headed by married couples in 2006—but these were not, by any measure, the nuclear family of the sitcoms. For instance, only 23% of U.S. householders that year were married and living with their own children (U.S. Census Bureau, 2008). This is related to the fact that around one third of all Americans are part of stepfamilies (Jones, 2003). A *stepfamily* is defined as a married-couple family household with at least one child under age 18 who is a stepchild (i.e., a son or daughter through marriage, but not by birth) of the householder (U.S. Census Bureau, 2010). Children in stepfamily arrangements are likely to go back and forth between parents for visitations or as part of shared custody.

Attitudes toward marriage have changed considerably. Fewer Americans are married today, either due to divorce or because they have never married at all. In 1970, 72% of adults in the U.S. were married, but by 1996 that had dropped to 60% (Kuttner, 2002). A significant number of single adults have never been married. The U.S. Census Bureau (2006) reported that in 2006, 42% of U.S. citizens age 18 or older were unmarried and 60% of unmarried adults had never been married.

Studies also show that people are marrying later. In 1960, the average age for a woman to marry for the first time was 20, while men first married around 23. By 2001 these average ages had risen to 25 and 27, respectively (Kantrowitz, et al., 2001). In addition, more-highly educated people are marrying even later than the national average and are having fewer children and divorcing less often (APA, 2004).

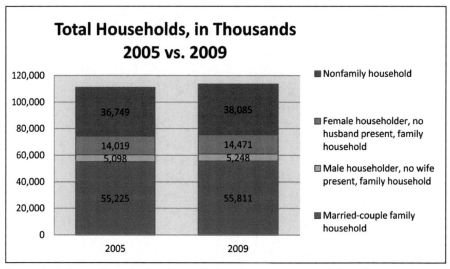

Source: U.S. Census Bureau, "Families and Living Arrangements," n.d., http://www.census.gov/
population/www/socdemo/hh-fam.html

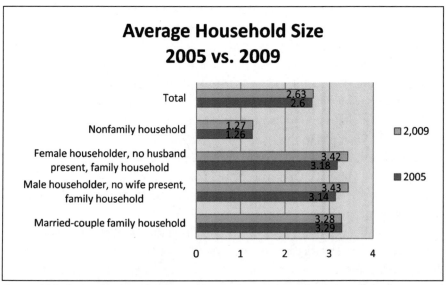

Source: U.S. Census Bureau, "Families and Living Arrangements," n.d., http://www.census.gov/
population/www/socdemo/hh-fam.html

The face of the American family has altered in other ways as well. Families themselves have gotten smaller over time. Family size for the nation as a whole has dropped from 3.29 in 1980 to 3.13 in 2006 (U.S. Census Bureau, 2008). Higher divorce rates and shifting marriage demographics have not only created more stepfamilies, but they have also dramatically increased the numbers of single-parent families, unmarried parents who may or may not cohabit, and grandparents who act as parents. Older generations are living longer; in 2000, households also included 5.5 million elders and disabled family members who needed care (APA, 2004). Changing laws and attitudes about homosexuality have also resulted in more openness about households with same-sex, joint householders who may or may not be married (depending on state laws) or have children. The U.S. Census Bureau (2008) reported that approximately 0.6% of U.S. households were headed by same-sex couples.

Rise in Single-Parent Families

The trend in single parenting has been on an upward trajectory since the middle of the 20th century. Between 1960 and 2002, unmarried households had increased by 72% (Kantrowitz et al., 2001, p. 46). Similarly, in 1990 only 12% of children lived with a single parent, but in just 6 years the rate more than doubled to 28% (Kuttner, 2002). This trend has persisted into the current century: In the year 2000, it was reported that 73% of households were headed by married couples (U.S. Census Bureau, 2008), but by 2006 that number had dropped to 71%. The percentage of households headed by single parents rose at the same rate during that time period. Households headed by single mothers between 2000 and 2006 increased from 22% to 23%, while the frequency that households would be headed by single fathers rose from 5% to 6% (U.S. Census Bureau, 2008). This increase was attributed not only to higher divorce rates, but also because more women were choosing to have children outside of marriage.

Another trend in single-parent households has been a steady increase in births to unmarried women over the last three decades. In 1980 only 18.4% of births were to women who were not married, but by 2003 34.6% of births were to unmarried women, and that percentage rose to 35.7% in 2004 (U.S. Census Bureau, 2008). Almost half of those women in 2003 and 2004 were living with men who may or may not have been the fathers of their children (Heuveline & Timberlake, 2004).

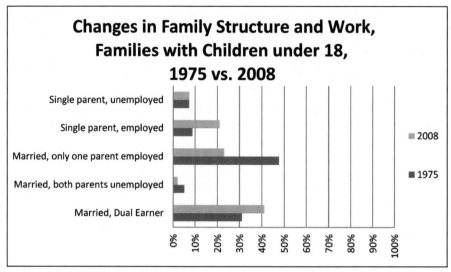

Source: U.S. Census Bureau, "Families and Living Arrangements," n.d., http://www.census.gov/population/www/socdemo/hh-fam.html

Today, single-parent-run households—which comprise at least 29% of all American households—face the same demands for shopping, household care, childcare, medical care, and possibly elder care as others. These parents need to pay the same bills as dual-income householders but from a single income. Single parents busy with child rearing and household matters also may find they have less time to keep up with technology or pursue higher education goals. Another important point to note is that these individuals are almost four times as likely to be women (U.S. Census Bureau, 2008).

Growth of Dual-Income Families

As noted, the number of dual-income families has been increasing since 1975. This is the result of more and more women joining or staying in the workforce with dependent children. A study by the American Psychological Association (APA) reports that 70% of women with school-aged children work outside the home. This includes 55% of mothers with infants (APA, 2004). Over half of mothers work or return to work while the children are very young, and up to 50% or more go to work once the kids are in school. Some mothers who remain home may work there, either by providing childcare or some other home-based self-employment.

Statistics also show that roles of men and women in the home are shifting to reflect changes in the workplace; however, they also indicate that women still carry a disproportionate share of these responsibilities. Pailhe and Solaz (2006) report that women who work as many hours as their spouses or partners, and even those who make more money, tend to spend considerably more time with housework and childcare than men, with the result that they are carrying an unequal share of the family burden. While working mothers spend an average of 2.5 hours a day on primary childcare, and 5.9 hours per day on primary and secondary childcare (which may include supervising and managing care for children), working fathers spend 1 hour and 2.3 hours performing such tasks, respectively. These numbers mean that women do over twice as much work in the home than men. It has been estimated that mothers also provide four times the amount of child-related travel and communication as fathers (Craig, 2006). One study shows that mothers in dual-earner households account for 76% of parental time with children (Pailhe & Solaz, 2006).

The amount of time that parents spend with their children also demonstrates another aspect to the evolution of gender roles in the home. Women spend more time with children, but men are increasing the time they spend with their children. Hook (2006) shows that in 1995, employed fathers spent roughly 1.5 to 2.5 hours per day with their children; in 2003, they were spending 6 hours a day with them. With all the changes in parental work and family schedules, children have roughly the same amount of time with parents as they have had in the last 10 or 15 years. A higher proportion of that time seems to be now with their fathers (APA, 2004).

What may have suffered in this complicated act of juggling schedules is the amount of time that families as a whole spend together. Children share the company of *both* parents together for an average of only 8 minutes per day (Pailhe & Solaz, 2006). This would also mean that the traditional TV-sitcom family dinner, during which all the family members talk about their day, is a thing of the past in many modern households. Family members spend time with each other separately or in small groups, but the whole family rarely has time to be together.

Increase in Military Families

Military families represent an important segment of our society in the context of change and working families. In 2006, according to Department of Defense statistics (as reported by the Sloan Work and Family Research Network

[SWFRN]), 43% of all military personnel, or about one million, were parents. Periodic deployments of one spouse, or in some cases, both spouses at the same time, create disruptions in the care of children and dependents. Unlike mainstream families, military families confront special circumstances, including frequent moves, family separation, financial difficulties, underemployment, and risk of injury or death of military spouses. In addition, members of the National Guard and Reservists face sudden deployments that require advance financial and legal planning to ensure care for their family and other personal matters (SWFRN, 2009a, 2009b).

While federal and state laws have been enacted in recent years to support military families with children, many challenges still exist, such as impact of deployment on child-care arrangements and problems related to transferring children to a new school. Military wives also are less likely to be employed than wives of civilians, and those who do work earn significantly less than their civilian counterparts (SWFRN, 2009a, 2009b).

A return to civilian life after long periods away presents new problems for military spouses as they struggle to pick up where they left off—be it a college education, a career, or household chores and responsibilities. Many find that returning to civilian life involves giving up the structured existence they were used to for a more loosely constructed, self-directed mode of living (University of Phoenix, 2010b). In the process, they may encounter trouble balancing life and work and reestablishing their roles as parents, spouses, and civilians at large.

Work and Family

The concept of linking work and family became common in the 1970s and 1980s as more women entered the professional workforce and sought quality care for their children. Large companies and corporations couldn't ignore women's needs to succeed professionally *and* have a family (Harrington, 2007). As a result, these organizations became more accommodating to the needs of working families, offering wellness programs, on-site childcare, availability of flextime and teleworking, and additional resources to help employees reach a work/life balance.

Today, work, family, health, and balance are important issues for many workers who are looking for fulfilling jobs and workplaces. According to Kathleen Gerson (2011, p. 399) the younger generation is looking for new ways to approach work, life, and family that don't require them to choose between spending time with their children and earning an income. Gerson also notes that both spouses should have jobs before marrying as an important step toward learning how to share responsibilities at home while holding jobs and careers.

Given the pressure on working parents, however, how do people actually perceive their ability to balance their home and work lives? In a study by Boston College's Center for Work and Family (2003), executives were asked to identify themselves as work-centric, family-centric, or both ("dual-centric") to determine their primary focus. The study revealed that 61% of the workers identified themselves as work-centric, while 32% identified themselves as dual-centric.

The Boston College researchers were particularly interested in learning more about dual-centric individuals and how their attitudes affected their work, home lives, and health. Within this category, an equal number of men and women identified themselves as dual-centric, and 62% of them had at least one child at home. Dual-centric individuals in the study worked about five hours less per week than their work-centric colleagues and were happier with their work. One interesting point was that in spite of working 5 hours less per week, the dual-centric workers felt more successful at work than those who self-identified as work-centric. (The ones with the lowest feelings of success were those at either extreme: very work- or very home-centric.)

Dual-centric workers also experienced considerably less stress in their work and in balancing their work and home lives. The same study also provided insights into gender differences in perception of career goals and success. Women executives were more likely than their male counterparts to have made important life decisions, like postponing marriage and delayed having children (Boston College Center for Work and Family, 2003).

Finding work/life balance is clearly an important consideration for most American families, who often face such very real issues as finding adequate childcare and jobs with flexible schedules and options for alternate work arrangements, both of which help individuals manage their lives in fulfilling ways. Employers are increasingly responding to the call for achieving parity of work and family time with unique options, from providing on-site childcare to offering wellness programs to promote health and well-being.

Childcare: Important to Parents and the Economy

Obtaining affordable, quality childcare, especially for children under the age of 5, is a major concern for many parents, especially as the numbers of two-parent working families have risen in recent years (U.S. Bureau of Labor Statistics [BLS], 2010). According to a study by the National Association of Child Care Resource & Referral Agencies (NACCRRA, 2010), more than 11 million children under the age of 5 in the U.S. are in some type of childcare arrangement every week. Children of working mothers spend an average of 36 hours a week in childcare. The trend in childcare assistance for employees has improved over the years. More employers (39%) were offering childcare assistance in 2008, up from 23% in 1998 (NACCRRA, 2010).

Childcare services in the U.S. is an important source of family income. Preschool teachers, teacher assistants, and childcare workers accounted for almost 78% of wage and salary earners in the child daycare industry in 2008. Child daycare services provided about 859,200 wage and salary jobs in 2008. There were an additional 428,500 self-employed and unpaid family workers in the industry, most of whom were family childcare providers, although some were self-employed managers of childcare centers (BLS, 2010).

Childcare options for working parents have seen gradual improvement over the past decade, but many barriers still exist. For instance, many American workers are eligible for only two weeks of vacation a year—the least in the industrialized world—and many American workers don't even take that small amount of time off from work (*Overwork in America*; Galinsky et al., 2004). Limited vacation time and workers' tendency not to take all available vacation days are important considerations in the context of childcare and school schedules. Students get out of school before the average workday ends, and when all of their holidays and summer vacations are added up, children are actually home from school for several months a year. These realities make effective childcare programs all the more essential.

While childcare services are more popular than ever, and their availability is rising, are they truly affordable? The annual cost of childcare can range from $4,560 to $18,733 for full-time services for an infant, and from $4,460 to $13,158 for a 4-year-old (NACCRRA, 2010). For professional and managerial workers who have good incomes, paying for childcare is not a major problem. For parents in lower income groups, however, affordable childcare can become a very real factor in the decision to take on additional work or home responsibilities. Furthermore, total expenditures and budgets for childcare and education have quadrupled since the 1960s.

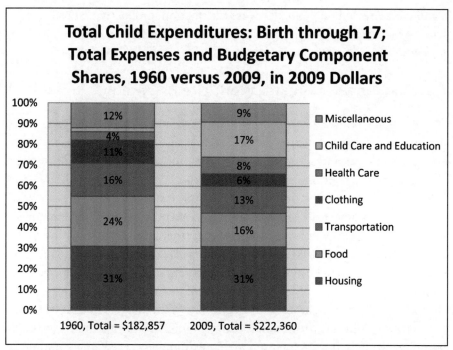

Total Child Expenditures: Birth through 17; Total Expenses and Budgetary Component Shares, 1960 versus 2009, in 2009 Dollars

1960, Total = $182,857   2009, Total = $222,360

Source: U.S. Department of Agriculture, Center for Nutrition Policy and Promotion, Expenditures on Children by Families, 2009, 2010, http://www.cnpp.usda.gov/publications/crc/crc2009.pdf

The role of employers in helping parents get access to childcare is increasing, though not at the same rate as employees' need for these services. A small percentage of employers provide on-site childcare or emergency care, and a few provide vouchers, subsidies, or other financial support. Most employers that provide help often do so by providing referrals or informational resources on childcare services (Galinsky & Bond, 1998; Hewitt Associates, 2003). Around 16% of employers allow employees to do some parenting in the workplace; this option, however, is more likely to be available to white-collar than to blue-collar workers (Secret, 2005).

Employers that do provide some form of childcare to their workers receive several benefits from this practice. One company found that without childcare services, an estimated 68% of employees would have had to miss some work (Bright Horizons, 2005) and that workers not only miss less work but are more positive about their jobs and less likely to quit when they have adequate childcare (Boston College Center for Work and Family, 2001). Because of childcare's significant effect upon employee work satisfaction, productivity, loyalty, stress levels—and, consequently, employee health—childcare is clearly a major issue for parents.

Flextime Provides Greater Work/Life Balance

One solution to balancing work and family responsibilities for single parents and dual-worker families is flextime: employees' ability to vary the times they start and stop work, take breaks midday or between shifts, or both. In addition, flextime programs sometimes permit workers to change their work hours from week to week, depending on other responsibilities.

While flextime work policies are often seen as an asset for employees, their importance to employers may rise over the coming years. In the global marketplace, employees often have to deal with customers in different countries, and thus need to be available during varying times of the day or week. An employee may confer with co-workers or clients in other time zones, early in the morning, late at night, or both in one day. With markets in countries such as India and China growing dramatically, this kind of time flexibility is rapidly becoming an important job requirement. Although people can communicate by email across time zones, personal contact and interaction still remain vital components of successful working relationships, and the ability to be flexible in work schedules is becoming the norm and a necessity.

The popularity and acceptance of flextime has been growing steadily over the past two decades. One study shows that its availability grew from 24% of U.S. workers to 31% between 1998 and 2008 (Galinsky, Bond, & Sakai, 2008). Similar trends prevail in Europe, where employers who insist on a fixed work schedule decreased from 71% in 1993 to 67% in 2003 (Anxo et al., 2006).

Small business owners and their employees are twice as likely to use flextime than workers with larger employers (Bond, Thompson, Galinsky, & Prottas, 2002). Since 97% of private U.S. employers have 500 or fewer employees (Small Business Administration, 2008), flextime has become a significant factor in the American workplace.

Differences in flextime usage among genders and age groups are also worth noting. In the United States, men are more likely to have flextime available in their jobs than women, but women are more likely to use flextime when they have it (Galinsky, Bond, & Hill, 2004). Age group usage also varies; men and women from Generation X are equally likely to use flextime if they can (DiNatale & Boraas, 2002).

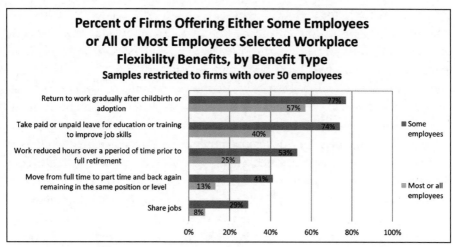

Percent of Firms Offering Either Some Employees or All or Most Employees Selected Workplace Flexibility Benefits, by Benefit Type

Samples restricted to firms with over 50 employees

Source: Executive Office of the President, Council of Economic Advisers, Work-Life Balance and the Economics of Workplace Flexibility, 2010, http://www.whitehouse.gov/files/documents/100331-cea-economics-workplace-flexibility.pdf

It is important to note, though, that flextime has proved very successful in the American work arena. People with access to it report such positive effects as job satisfaction, improved mental health, low levels of spillover from work to home life, and higher productivity (Bond et al., 2002). Moreover, 88% of low-wage employees and 87% of high-wage employees report that having the flexibility they need to manage work and personal or family life would be "extremely" or "very" important if they were looking for a new job (Bond & Galinsky, 2008).

Teleworking Brings Multiple Worker and Employer Benefits

A 2010 Telework Research Network report found that teleworking or telecommuting offers many economic, environmental, and social benefits. Advantages were perceived for employers in productivity, real estate savings, and lower turnover and absenteeism. Specifically, industry statistics cited in the study showed productivity improved on average between 20% and 35%. Dow Chemical estimated a 32.5% increase in productivity among its teleworkers, and IBM workers suggested its employees could be up to 50% more productive. Similarly, American Express stated that its telecommuters handled 26% more calls and produced 43% more business than their office-based counterparts (Lister & Harnish, 2010).

Employee benefits included savings in gas and work-related expenses, and reduced time spent in commuting. Researchers also observed that teleworkers may remain healthier because they are less exposed to sick co-workers, occupational and environmental hazards, and the daily commute—possibly resulting in lower medical costs. The study also showed community benefits through less gasoline usage, lower greenhouse gas emissions, savings on highway maintenance, and a reduced number of accidents.

Several companies have developed programs to encourage their workers to telework part-time. In 2009, Cisco Systems conducted a worldwide teleworker survey of its employees and found that the average teleworker worked from home two days a week. The majority of these workers reported that teleworking gave them flexibility in both their work and personal lives, increased productivity, and that they believed it helped the environment. Based on the survey results, Cisco calculated that its teleworking employees worldwide prevented 47,320 metric tons of greenhouse gases from being released into the environment in 2008 and that teleworking saved those employees $10.3 million in fuel costs (Cisco Systems, 2009).

The trend in teleworking is most noticeable among self-employed workers, who are three times more likely to work at home than other employees (BLS, 2009b). Small business owners are also considerably more likely to work from home (Bond et al., 2002). Employees who can work more hours from home include a broad cross-section of professionals whose work requires only a computer (e.g., writers, graphic designers, and accountants) to employees in the personal-care services, such as childcare and elder care (BLS, 2009b). Professional and managerial workers are also seeing more opportunities to telework, as are many people in the technical, sales, and administrative support areas.

In one study (Bond et al., 2002), teleworkers were found to experience less fatigue, feel less drained, and experience more satisfaction than workers without this option. Half of the workers with flexible hours reported high levels of life satisfaction, which is consistent with the results of the Cisco survey. Teleworking also provides reasonable accommodations for workers with disabilities (Richman, Noble, & Johnson, 2002). In contrast, working poor families—in which both parents may have to work two jobs to pay the bills—rarely have the option of telecommuting, as they often work in the service sector or outside typical business hours (BLS, 2009b).

A major impetus to the rise in telecommuting has been the increased availability and use of broadband Internet access. This use grew rapidly up until

2009 before peaking and leveling out. The demographics of people who have broadband Internet access, however, also reflect social differences in the population as a whole and may predict the segment of people who can take advantage of the teleworking option. Men (61%) are more likely to have broadband access than women (58%); Whites (63%) have more access than either non-Hispanic Blacks (52%) or Hispanics (47%). People with more money and/or more education, and who are urban residents, are all more likely to have access to and use broadband.

Teleworking also has the potential to help parents and other householders balance the demands of home and work. If people can work from home or from the doctor's office, from dance lessons or basketball practice, they will have a lot more flexibility in their schedules, which may translate into less stress in their lives, more time with their families, lower childcare costs, and fewer healthcare bills for stress-related complaints.

The Networked Family

In a 2008 Pew Research Center study (Kennedy, Smith, Wells, & Wellman, 2008), American families were found to use a wide range of communications media to stay in contact with each other. Married couples—in particular, those with young children—more frequently used the Internet and cell phones and were more likely to use computers and broadband access than other family units. The authors note that while many fear that technology pulls families apart, they found that it is actually enabling a new kind of connectedness. Cell phones, for example, help couples connect and coordinate their lives by keeping them in contact with each other all day, instead of catching up at home after work as in the past.

Other benefits noted include sharing moments with each other in real time while they are happening. Pew noted that 47% of married couples contact each other once a day and that they are increasing their use of cell phones versus land lines. Owning a cell phone also increases the frequency of communication between couples. The Pew data follow the general perception among families that technology allows their family lives to be as close as or closer than what they experienced while they were young.

- 70% of couples who both own a cell phone contact each other once a day or more to say hello or chat; 54% of couples who have one or no cell phones do this at least once a day.

- 64% of couples who both own a cell phone contact each other at least once a day to coordinate their schedules; 47% of couples who have one or no cell phones do this at least once a day.

- Parent–child communications, particularly on a daily basis, are similarly dominated by the telephone: 42% of parents contact their offspring daily using a cell phone, and 35% do so using a landline telephone (Kennedy et al., 2008).

The survey further discusses the impact of technology on friends and co-workers. A third of adults surveyed note that the Internet has improved their connection to friends "a lot," and 23% say it has increased the quality of family-member communication.

Growth of Corporate Wellness Programs

In their efforts to retain the best-trained and most talented workers, employers now offer wellness programs that promote employee health and fitness. This twofold strategy is designed to decrease the work time lost from employee illness while lowering their healthcare costs.

It has been estimated that 90% of U.S. employers provide some kind of wellness program, but what they consider to be a "program" may involve simply providing health information and may not be very effective (Parks & Steelman, 2008). Comprehensive wellness programs have been defined as "health-oriented programs designed to maintain a level of well-being in workers through proper diet, exercise, stress reduction, disease management, and illness prevention" (Kopicki, Van Horn, & Zukin, 2009, 1).

An estimated 45% of multinational companies had wellness programs in place in 2008, and an additional 16% were planning them (Slutzky, 2008). Many large companies consider wellness an important measure of their success in employee retention and overall satisfaction. SAS, one of the world's largest privately owned software companies, provides a free 66,000-square-foot fitness center and swimming pool, a lending library, and a summer camp for children. The company is also said to have the lowest turnover rate in the industry, at 2% (CNN Money, 2011).

Many companies also offer a variety of wellness packages for working families (WorkingMother.com, 2010). They include discounted rates for childcare (Abbott Laboratories, 2011), emergency at-home care at a nominal charge (Accenture), on-site medical screening and mammograms (Microsoft), and on-site massage therapy and fitness classes (Google.com). Facebook pays 100% of employee benefit premiums and 50% of any dependent premiums in the United States (facebook.com).

The results of wellness programs in general have been impressive, according to recent studies. The benefits include reduced absenteeism, improved morale, and increased employee loyalty (Kondracki, 2008; Parks & Steelman, 2008), and the programs have saved employers significant amounts of healthcare costs. The U.S. Centers for Disease Control and Prevention estimates that each dollar invested by a company in a wellness program saves the employer $3.50 in healthcare and health-related costs. It also reduces absenteeism, workers' compensation claims, and healthcare costs by 28%, 30%, and 26%, respectively (Hurst, 2010). Similarly, the U.S. Department for Health and Human Services estimated that in 2003 each dollar that employers spent on wellness programs brought between $1.49 and $4.91 as a return on investment, for a median of $3.14 (Ickes & Sharma, 2009).

Wellness Programs Offer a Prescription for Savings
The National Business Group in Health, a Washington, DC–based organization that specializes in corporate healthcare issues, has analyzed current data that indicate companies implementing comprehensive wellness programs see not only reduced overall healthcare expenditures but also a reduction in year-over-year health insurance premium increases when compared with companies that do not offer them.

[Francis, R. (2009, November 9–15). Corporate wellness programs show strength in downturn. *Fort Worth Business Press*, p. 1/21.]

Workers associate employer-sponsored wellness programs with better health. About 30% of workers in one survey reported that wellness programs have a major impact on the health of people in the workplace, while a little over half of the respondents reported a minor impact. Overall, 70% of workers thought employers should provide wellness programs, and 80% of workers who already have access to a wellness program thought so (Kopicki, Van Horn, & Zukin, 2009).

Workers in large companies (250 or more employees) and those who are more educated and affluent are more likely to have access to comprehensive wellness programs in the workplace. Over twice as many workers with incomes of $70,000 or more have access to wellness programs as those making $35,000 or less (44% vs. 21%). Forty-six percent of workers with a college degree reported they had access to a wellness program at work, versus 25% of people with a high school education. Salaried workers also had more access than hourly workers (45% vs. 35%; Kopicki et al., 2009).

Though programs are not yet equally available to all workers, health issues have become so important that wellness is fairly likely to remain an important part of the workplace for the foreseeable future.

Impact of Family Trends on Higher Education

Shifting patterns in American family life and the very real ways in which these changes are impacting the workplace have also produced changes in the American higher education system. With such profound transformations, it's worth inquiring into the makeup of today's college and university students. Are higher education programs adapting to today's family structures, work environments, and social demographics?

Nontraditional Students

Back in the *Leave It to Beaver* days, the average college student was male; he attended college straight out of high school, lived on campus, and received financial aid from his parents. He might have worked part-time, but more likely he focused full time on his education and extracurricular activities.

Now, those days have passed. Just as the nuclear-family model of yesteryear is fading, this older image of college students has become obsolete. In fact, the average student today is often seen as a young man or woman who attends college right after high school, lives on campus, and gets financial help from his or her parents. In reality, these traditional students make up only 27% of undergraduates, who are primarily served by four-year institutions with restrictive admission policies that have been tailored to their demographics and financial-aid needs (University of Phoenix, 2010a).

The majority of today's undergraduates are *nontraditional* students, who comprise a heterogeneous mix of working individuals (full- or part-time), parents (married or single), stop-outs (people who have taken a break for any number of reasons), veterans, the economically disenfranchised, and those who are financially independent from parents. The U.S. Department of Education identifies nontraditional students as those likely to be in their 20s, work while attending school, and raising children. The traits that make these students nontraditional also place them statistically at risk of not completing their educational programs (Brock, 2010, p. 113).

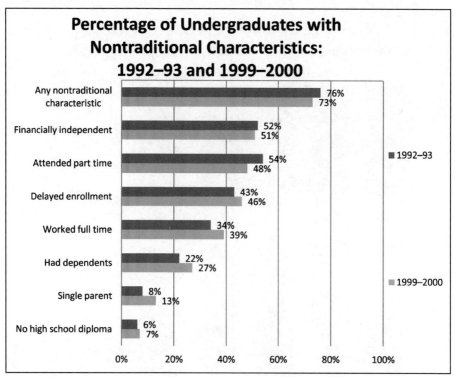

Percentage of Undergraduates with Nontraditional Characteristics: 1992–93 and 1999–2000

- Any nontraditional characteristic: 76% (1992–93), 73% (1999–2000)
- Financially independent: 52% (1992–93), 51% (1999–2000)
- Attended part time: 54% (1992–93), 48% (1999–2000)
- Delayed enrollment: 43% (1992–93), 46% (1999–2000)
- Worked full time: 34% (1992–93), 39% (1999–2000)
- Had dependents: 22% (1992–93), 27% (1999–2000)
- Single parent: 8% (1992–93), 13% (1999–2000)
- No high school diploma: 6% (1992–93), 7% (1999–2000)

Source: National Center for Education Statistics, "The Condition of Education: Closer Look 2002a, Nontraditional Undergraduates," n.d., http://nces.ed.gov/programs/coe/analysis/2002a-sa01.asp

Ethnicity, gender, and diversity are also important factors in the overall composition of higher education students. It is predicted that by 2020, minority undergraduates will outnumber White students in the United States. In 1970, when the federal government first began to report the demographics of college students, the majority were male. By 2005, that gender ratio had shifted and the majority were now female. The ratio has essentially reversed, with female students making up the same percentage male students did in 1970 (Brock, 2010).

Between 1965 and 2005 college enrollment increased nearly 300%, which included higher enrollment for all minority groups. Success rates, however, have failed to follow this trend. Larger numbers of students in all racial groups are graduating, but larger numbers are also failing to complete their programs (Brock, 2010).

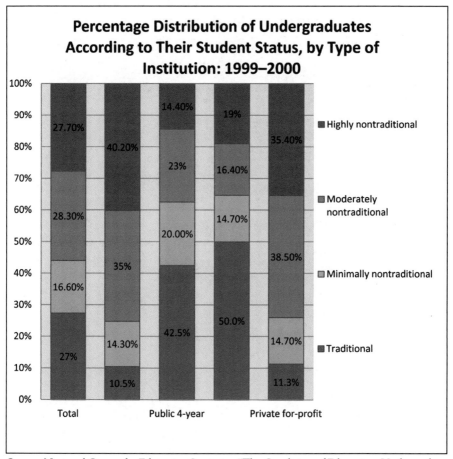

Source: National Center for Education Statistics, "The Condition of Education: Undergradute Enrollment (Indicator 8–2011)," n.d., http://nces.ed.gov/programs/coe/indicator_hep.asp

Women and men have somewhat different persistence patterns in completing programs as well. Female students are more likely to complete four-year programs than male students are, and male students are more likely to complete two-year programs than female students. Because considerably more women enroll in both kinds of programs, more women than men receive degrees (Brock, 2010). One more fact influencing success and preparedness is that American colleges are finding that high schools are graduating students without the academic skills needed to succeed at the postsecondary level, leaving them to pick up the slack (University of Phoenix, 2010a).

Finally, immigrant groups, as well as native-born minorities, are attending higher-education programs in significant numbers. Among American-born

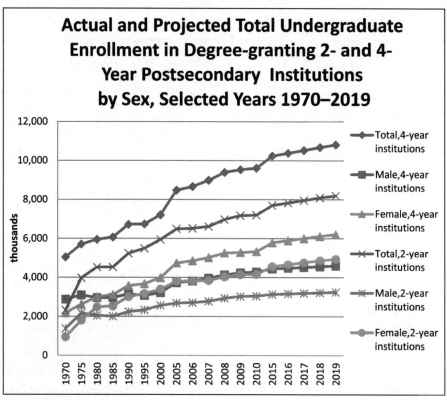

Actual and Projected Total Undergraduate Enrollment in Degree-granting 2- and 4-Year Postsecondary Institutions by Sex, Selected Years 1970–2019

Source: National Center for Educational Statistics, "Projections of Education Statistics to 2019, Table A-19," 2010, http://nces.ed.gov/programs/projections/projections2019/tables/table _A19.asp?referrer=list

students, 17.6% complete a bachelor's degree, while 16.1% of immigrant students do so. In some immigrant groups this percentage is much higher. For example, 29.2% of students from South and East Asia and 25.6% of those from the Middle East complete bachelor's degrees (Pew Hispanic Center, 2008).

Nontraditional Education Programs

As stated earlier, higher education enrollment has grown exponentially since 1965. While traditional students are much more likely to enroll in a four-year college program, there has also been a distinct shift in enrollment in nontraditional programs. In 1969, 26% of U.S. students were attending two-year institutions, but by 2005 that had increased to 37%. In addition, the trend shows that female, black, and Hispanic students are disproportionately enrolled in community colleges (Brock, 2010).

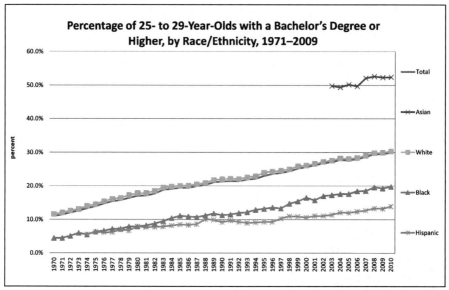

Source: National Center for Education Statistics, "The Condition of Education: Educational Attainment (Indicator 24–2011)," n.d., http://nces.ed.gov/programs/coe/indicator_eda.asp

Another statistic of note is that while enrollment in traditional college programs grew by 1.2% from 2008 to 2009, online enrollments grew by 17% during the same period (Allen & Seaman, 2010). The demographic makeup of students seeking education is also changing. As of 2004, 43% of community college students and 12% of four-year college/university students were adult learners. In addition, 40% percent of all college and university students were 25 or older (*Access to Higher Education*, 2009).

When an increasing number of enrolled students are balancing their education with work, past educational models are no longer suitable. Fewer students have the time and resources required to attend classes as residents, and more expect educational programs to adapt to their changing needs and circumstances. To some extent, universities are becoming more cognizant of this fact but not as rapidly as they should. Technology changes society swiftly, but many universities are responding to these transformations at a glacial pace.

There are, however, a few nontraditional higher education programs that are able to meet the challenge of educating this majority of students who do not fit the traditional model. The post-9/11 GI Bill, for instance, has helped many veterans return to college and resume their education and training (Sewell, 2010). The open-admission policy of many online educational institutions is particularly attractive to veterans, who consistently score higher on literacy competencies and often are able to complete their academic programs as well (University of Phoenix, 2010b). The University of Phoenix serves a broad cross-section of veterans, of whom 29% are African American and 13% are Hispanic. Additionally, it serves a larger percentage of minorities and women than the national average. In 2009, more than 50% of its graduate students were minorities, compared with a national average of 36%. Sixty-seven percent of undergraduates that year were women, while 69% of graduate students were women, as compared to the national averages of 57% and 60%, respectively (University of Phoenix, 2009).

Real needs for highly educated workers are also being felt in many of today's industries, most notably in the manufacturing and high-tech sector, where companies seek employees with technical backgrounds and knowledge of global markets. In 2009, the BLS (2009a) reported that one third of the projected 15.3 million job openings during the next decade will require a postsecondary degree. Currently, the number of students with the ability and resources to attend college full time directly out of high school may not be large enough to meet the future needs of employers for degree-equipped workers. Similarly, with constant changes in technology and the job market, a single training program may not be sufficient for a worker's entire career. They will need to continue to learn throughout their working lives to keep abreast of changes and new information about the global village in which they live.

There is an important financial consideration as well. An adult with a bachelor's degree will earn approximately one third more over the course of a lifetime than someone who does not complete college. She or he will earn about twice as much as someone with only a high school education (Brock, 2010).

These changes in the educational landscape mean that higher education organizations must seriously address the needs of nontraditional students by developing programs for a cadre of lifelong learners. Today's 21st-century workplace is more mobile and flexible and is changing at a rapid clip. Learning environments must adapt accordingly.

Summary

An understanding of the modern family is integral to gaining deeper insights into how higher education must change to keep up with the times. The traditional American nuclear family of previous generations has changed dramatically. The classical model of a single-wage family consisting of a husband, wife, and two or three children is the norm no more. Today's society consists of a host of optional models, such as single parents with many children, same-gender families with adopted children, dual-wage, or military families—all working diligently to balance work and life issues. Important needs among these different family structures include adequate childcare, flextime and teleworking options, and health and wellness programs for employees and their families. As individuals manage their personal family dynamics and strive to be productive in society, they expect higher education institutions to provide flexible programs, schedules, and options for learning and engagement.

References

Abbott Laboratories. (2011). Benefits. Retrieved from http://www.abbott.com/global/url/content/en_US/50.30.30:30/general_content/General_Content_00049.htm

Access to higher education for the adult learner. (2009). Retrieved from http://www.womeningovernment.org/files/file/higher-ed/toolkit/AccesstoHigherEducationfortheAdultLearner Presentation.pdf

Allen, I. E., & Seaman, J. (2010). *Learning on demand: Online education in the United States, 2009.* Retrieved from http://www.sloan-c.org/publications/survey/pdf/learningondemand.pdf

American Psychological Association. (2004). *Public policy, work, and families: The report of the APA presidential initiative on work and families.* Retrieved from http://www.apa.org/pubs/info/reports/work-family.aspx

Anxo, D., Boulin, J. Y., Fagan, C., Cebrián, I., Keuzenkamp, S., Klammer, U., . . . Toharía, L. (2006). *Working time options over the life course: New work patterns and company strategies.* Retrieved from the European Foundation for the Improvement of Living and Working Conditions website: http://www.eurofound.europa.eu/pubdocs/2005/160/en/1/ef05160en .pdf

Bond, J. T., & Galinsky, E. (2008). *2008 National Study of the Changing Workforce. Workplace flexibility and low wage employees.* Retrieved from Families and Work Institute website: http://www.familiesandwork.org/site/research/reports/WorkFlexAndLowWageEmployees.pdf

Bond, J. T., Thompson, C., Galinsky, E., & Prottas, D. (2002). *Highlights of the national study of the changing workforce.* Retrieved from Families and Work Institute website: http://familiesandwork.org/site/research/summary/nscw2002summ.pdf

Boston College Center for Work and Family. (2001). *Bristol-Myers Squibb on-site child care center assessment* (final report). Chestnut Hill, MA: Author.

Boston College Center for Work and Family. (2003). *Leaders in a global economy: A study of executive men and women* (executive summary). Retrieved from http://www.bc.edu/content/dam/files/centers/cwf/research/highlights/pdf/GlobalLeadersExecutiveSummary.pdf

Bright Horizons Inc. (2005). *CIBC children's center.* Watertown, MA: Bright Horizons Family Solutions. Retrieved from http://wfnetwork.bc.edu/encyclopedia_template.php?id =4331&topic=29&linktype=overview&area=All

Brock, T. (2010, Spring). Young adults and higher education: Barriers and breakthroughs to success. *The Future of Children, 20*(1), 109–132.

Chinhui, J., & Potter, S. (2006). Changes in labor force participation in the United States. *Journal of Economic Perspectives, 20*(3), 27–46. Retrieved from EBSCOhost.

Cisco Systems. (2009). Cisco study finds telecommuting significantly increases employee productivity, work-life flexibility and job satisfaction [Press release]. Retrieved from http://newsroom.cisco.com/press-release-content?type=webcontent&articleId=5000107

CNN Money. (2011). 100 best companies to work for. Retrieved from http://money.cnn.com/magazines/fortune/bestcompanies/2011/index.html

Craig, L. (2006). Does father care mean fathers share? A comparison of how mothers and fathers in intact families spend time with children. *Gender and Society, 20*, 259–281.

DiNatale, M., & Boraas, S. (2002). The labor force experience of women from "Generation X." *Monthly Labor Review, 125*(3), 3–15.

Fields, J. (2004, November). *America's families and living arrangements: 2003. Population characteristics*. Retrieved from U.S. Census Bureau website: http://www.census.gov/prod/2004pubs/p20-553.pdf

Galinsky, E., & Bond, J. T. (1998). *Business work-life study: A sourcebook*. New York: Families and Work Institute.

Galinsky, E., Bond, J. T., & Hill, E. J. (2004). *When work works: A project on workplace effectiveness and workplace flexibility*. Retrieved from http://www.familiesandwork.org/3w/research/status.html

Galinsky, E., Bond, J. T., Kim, S. S., Backon, L., Brownfield, E., & Sakai, K. (2004).*Overwork in America: When the way we work becomes too much*. Retrieved from Families and Work Institute website: http://familiesandwork.org/summary/overwork2005.pdf

Galinsky, E., Bond, J. T., & Sakai, K. (2008). *2008 National Study of Employers*. Retrieved from Families and Work Institute website: http://familiesandwork.org/site/research/reports/2008nse.pdf

Gerson, K. (2011). Moral dilemmas, moral strategies, and the transformation of gender. In J. Z. Spade & C. G. Valentine (Eds.), *Kaleidoscope of gender: Prisms, patterns, and possibilities* (3rd ed., pp. 398–406). Newbury Park, CA: Pine Forge Press.

Harrington, B. (2007). *The work-life evolution study*. Retrieved from Boston College Center for Work & Family website: http://www.bc.edu/content/dam/files/centers/cwf/pdf/Work_Life_Evolution_Study_final.pdf

Heuveline, P., & Timberlake, J. M. (2004, December). The role of cohabitation in family formation: The United States in comparative perspective. *Journal of Marriage and Family, 66*, 1214 1230.

Hewitt Associates. (2003). *SpecSummary: United States salaried work/life benefits, 2003–2004*. Lincolnshire, IL: Hewitt Associates.

Hook, J. L. (2006). Care in context: Men's unpaid work in 20 countries, 1965 2003. *American Sociological Review, 71*, 639–660.

Hurst, E. F. (2010, February 15–21). Making health insurance more affordable for small employers: Use wellness programs to prevent costly problems. *The Enterprise* (Salt Lake City).

Ickes, M., & Sharma, M. (2009). Worksite health promotion: A practical strategy for obesity prevention. *American Journal of Health Studies, 24*(3), 343–52.

Jones, A. C. (2003, April 1). Reconstructing the stepfamily: Old myths, new stories. *Social Work*. Retrieved from http://www.highbeam.com/library/doc3.asp?docid=1G1:100767739

Kantrowitz, B., Wingert, P., Scelfo, J., Springen, K., Figueroa, A., Brant, M., & Abrams, S. (2001, May 28). Unmarried, with children. *Newsweek*, 46.

Kennedy, T. L. M., Smith, A., Wells, A. T., & Wellman, B. (2008). *Networked families*. Retrieved from http://pewresearch.org/pubs/998/networked-families

Kondracki, K. (2008, February 15). Worksite-wellness programs save more than just money. *The Central New York Business Journal*, 11.

Kopicki, A., Van Horn, C., & Zukin, C. (2009). *Healthy at work? Unequal access to employer wellness programs*. New Brunswick, NJ: John J. Heldrich Center for Workforce Development, Edward J. Bloustein School of Planning and Public Policy.

Kuttner, R. (2002, April 28). The politics of family. *The American Prospect, 13*(7), 22–23.

Lister, K., & Harnish, T. (2010). *Workshifting benefits: The bottom line*. Retrieved from TeleworkResearchNetwork.com website: http://www.workshifting.com/downloads/downloads/Workshifting%20Benefits-The%20Bottom%20Line.pdf

National Association of Child Care Resource & Referral Agencies. (2010). *Child care in America: 2010 State fact sheets*. Retrieved from http://www.naccrra.org/publications/naccrra-publications/publications/State_Fact_Bk_2010_sect01_070710.pdf

Pailhe, A., & Solaz, A. (2006). Time with children: Do fathers and mothers replace each other when one parent is unemployed? *European Journal of Population, 24*, 211–236.

Parks, K. M., & Steelman, L. A. (2008). Organizational wellness programs: A meta-analysis. *Journal of Occupational Health Psychology, 13*(1), 58–68.

Pew Hispanic Center. (2008). *Statistical portrait of the foreign-born population in the United States, 2008*. Retrieved from http://pewhispanic.org/factsheets/factsheet.php?FactsheetID=59

Richman, A., Noble, K., & Johnson, A. (2002). *When the workplace is many places. The extent and nature of off-site work today*. Watertown, MA: WFD Consulting.

Secret, M. (2005). Parenting in the workplace: Child care options for consideration. *Journal of Applied Behavioral Science, 41*, 326–347.

Sewall, M. (2010, June 13). Veterans use new GI bill largely at for-profit and 2-year colleges. *The Chronicle of Higher Education*. Retrieved from http://chronicle.com/article/Veterans-Use-Benefits-of-New/65914/

Shriver, M., & The Center for American Progress. (2009). *The Shriver report: A woman's nation changes everything* (H. Boushey & A. O'Leary, Eds.). Retrieved from http://www.americanprogress.org/issues/2009/10/womans_nation.html

Sloan Work and Family Research Network. (2009a, June). *Effective workplace series.Work-family information on: Military families*. Retrieved from http://wfnetwork.bc.edu/pdfs/EWS14_militaryfamilies.pdf

Sloan Work and Family Research Network. (2009b, June). *Policy mini-brief series. Work-family information on: Military families*. Retrieved from http://wfnetwork.bc.edu/pdfs/minib_military_families.pdf

University of Phoenix. (2009). Academic annual report. Retrieved from http://www.phoenix.edu/about_us/publications/academic-annual-report/2009.html

University of Phoenix. (2010a). *Extraordinary committment: Challenges and achievements of today's working learner*. Phoenix, AZ: Author. Retrieved from http://cdn-static.phoenix.edu/con-

tent/dam/altcloud/doc/extraordinary-commitment.pdf?cm_sp=UOPX+Knowledge+
Network-_-PDF-_-Extraodinary+Commitment

University of Phoenix. (2010b). *The face of today's military student*. Retrieved from
http://www.phoenix.edu/uopx-knowledge-network/articles/working-learners/enlisted-military-in-higher-education.html

U.S. Bureau of Labor Statistics. (2009a). Employment projections: 2008–2018 summary [Press release]. Retrieved from http://www.bls.gov/news.release/ecopro.nr0.htm

U.S. Bureau of Labor Statistics. (2009b). Work-at-home patterns by occupation. *Issues in Labor Statistics, 9*(2). Retrieved from http://www.bls.gov/opub/ils/pdf/opbils72.pdf

U.S. Bureau of Labor and Statistics. (2010). *Career guide to industries, 2010–11 edition: Child day care services. Nature of the industry*. Retrieved from http://www.bls.gov/oco/cg/cgs032.htm#nature

U.S. Census Bureau. (2006). *Statistical abstract of the United States: 2003. Section 2. Vital statistics*. Retrieved from http://www.census.gov/prod/2004pubs/03statab/vitstat.pdf

U.S. Census Bureau. (2008). *Statistical abstract of the United States: 2008 edition*. Retrieved from http://www.census.gov/compendia/statab/2008/2008edition.html

U.S. Census Bureau. (2010). Housing and Household Economic Statistics Division. Retrieved from http://www.census.gov/population/www/cps/cpsdef.html

U.S. Small Business Administration, Office of Advocacy. (2008). *The small-business economy: A report to the President*. Retrieved from http://archive.sba.gov/advo/research/sb_econ2008.pdf

WorkingMother.com. (2010). 2010 Working Mother 100 best companies. Retrieved from http://www.workingmother.com/node/3836/list

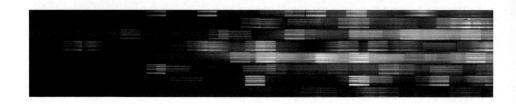

· 2 ·

NEW LEARNERS

Say the words "college student," and most people will picture an 18- to 22-year-old who lives on campus, attends school full-time, and is supported by his or her parents. That description, however, only applies to 27% of students enrolled in higher education today. The majority of "college students"—some 73%—are nontraditional students, also known as working learners (Choy, 2002).

Defined as students who are 23 years of age or older, finance their own education, and work part- or full-time while attending classes, working learners span every income and age bracket, come from every ethnic group, and work in a variety of industries, from retail and healthcare to business and manufacturing. Working learners have varying life circumstances and motivations for returning to school: He or she might be a single parent hoping to move out of a low-paying service-sector job by earning an associate's degree, a businessperson seeking an MBA to improve his or her chances for promotion, a career changer switching job fields or starting a business in midlife, or a highly placed executive performing research as part of a doctoral dissertation.

Working learners have different needs than traditional college students. Work and family responsibilities make heavy demands on their time, and attending classes scheduled during weekdays may be difficult. Forty-five percent of students enrolled in four-year colleges work 20 or more hours per week; among community college students, this percentage rises to 60% (Johnson,

Rochkind, Ott, & DuPont, 2009). Working learners also must cope with life events that most younger people have not yet faced, such as marriage or divorce, having children, losing a job, becoming ill, or suddenly having to care for a sick family member—all of which can interrupt or curtail their educational careers. If they have dependents—as do 23% of American students—they also may need to arrange childcare while they work and attend classes (Johnson et al., 2009).

Most institutions of higher education, however, are still designed to suit younger students without dependents and who work part-time or not at all. To better serve the large and growing working learner population, educators need to better understand these students' needs and the challenges they face when pursuing higher education.

Types of Working Learners

Working learners are a diverse population but one with many distinct subgroups that have different needs and requirements. Some of these subgroups are described below.

First-Generation Students

About 1 out of every 6 new students enrolled at institutions of higher education is a first-generation college student: a student whose parents never attended college (Capriccioso, 2006). Many first-generation students are also working learners who finance all or part of their education. These students are more likely than second-generation students to delay their entry into college until they have been in the workforce for several years. Many are minorities, immigrants, or the children of immigrants, and many are also of low socioeconomic status (Tym, McMillion, Barone, &Webster, 2004).

A significant barrier to first-generation students' academic success is their lack of cultural capital or knowledge about the ways and mores of higher education. Second- and third-generation students learn from friends, family members, teachers, and guidance counselors about how higher education works. They know how to choose a college, write an admittance essay, score well on college entrance exams, apply for financial aid, select classes and majors, and perform other tasks that are part of higher education. These students typically have grown up with the expectation that they will graduate from college, and they have attended good K-12 schools that prepared them well for college-level academics (Pascarella, Pierson, Wolniak, & Terenzini, 2004).

Many first-generation students lack these advantages. They may need extra guidance in procedures like applying to college, selecting classes, and pursuing financial aid. They may also have attended subpar K-12 schools and need remedial classes before they can achieve at the same level as their peers (Pike & Kuh, 2005). Due to their weaker academic preparation, first-generation students may need to learn such skills as studying for exams, using the library, performing research, managing their time, and writing long papers (Tym et al., 2004).

Reentry Students

Each year, many students drop out of higher education programs. In fact, over the past decade, only 57% of undergraduates completed bachelor's degree programs within 6 years (NCES, 2009). Students do not drop out only or even mainly for academic reasons. Life events, such as the serious illness of family members or one's self, job losses, pregnancies, births, marriages, divorces, and moves, can interfere with a person's education. Due to their work and family circumstances, working learners are more vulnerable to these disruptive events (Johnson et al., 2009). However, many students who leave higher education return to complete their degrees once their life situations stabilize. These reentry students may need remediation or refreshing of academic skills that have weakened in their time away from school.

Real Costs (Other Than Tuition) That Working Mothers Consider When Choosing Degree Programs

Cost of food when "eating on the run"

Gasoline & automobile repairs

Home health aides & eldercare staff

Lost wages from shortened work days

Books & course materials

Child day care & after-school care

Parking permits & toll-booth fees

Source: University of Phoenix Knowledge Network. (2010). *Extraordinary commitment: Challenges and achievements of today's working learner.* Phoenix, AZ: Author.

Single Mothers

The majority of working learners are women. Many female working learners also have children, and many are raising children on their own (NCES, 2002). At the University of Phoenix, one of the largest providers of education to working learners, for example, 65% of bachelor's degree students are women and 38% of those women are single mothers (University of Phoenix, 2009). Earning a degree can be especially challenging for single mothers, as many live below the poverty line and hold low-paying service-sector jobs that may not pay a living wage even for full-time work. Such jobs often involve irregular scheduling and strict policies against lateness and sick days, which can make it difficult to schedule coursework and other tasks associated with education (e.g., registration, buying books, and seeking financial aid; Quinn & Allen, 1989).

Military Personnel and Retired Military

An estimated 1.5 million Iraq and Afghanistan veterans are expected to finance higher education through the Post 9/11 Veterans Educational Assistance Act of 2008, popularly known as the "new GI bill" (American Council on Education, 2008). Like working learners in general, military students tend to have dependents and be older than traditional-age college students. Many are also first-generation college students (American Council on Education, 2008). Military students typically pursue higher education for one of two reasons: to increase their chances of promotion within the military or to prepare for civilian careers after they retire from military service.

Attending school as an active member of the military can be difficult. Military students often take longer to finish degree programs because they move frequently, requiring them to leave one school and enroll in another. The credits they earn do not always transfer from one school to the next, nor do schools always grant credit for knowledge or experience gained in the military (Cook & Kim, 2009). Students in the armed forces may also be deployed at any time, forcing them either to drop out of school or take a leave of absence. These interruptions can cause them to lose focus upon their academic careers, and many military students drop out of programs without earning degrees (DiRamio, Ackerman, & Mitchell, 2008).

Career Advancers and Career Changers

A significant number of working learners already have one or more degrees but are pursuing further education to improve their career prospects. Examples of this type of student include nurses with two-year licensed practical nurse degrees who want to become bachelors of science in nursing and executives with bachelor's degrees seeking to increase their skill sets by achieving MBAs. Other working learners choose to pursue new career paths by earning degrees in different fields from the ones in which they currently work. For example, professionals with bachelor's degrees who hold jobs in business or engineering might seek master's degrees in education or nursing to switch to those fields. Such working learners are pragmatic and job centered. They want to apply their classroom learning in the workplace immediately and graduate within a reasonable length of time, without taking too many courses that are irrelevant to their jobs.

Researchers

Many working learners are pursuing doctoral or other terminal degrees in their field of study. Most frequently, these students are mid- or late-career professionals who have identified a problem on the job or in society that they wish to investigate through research. They enter doctoral programs to acquire advanced-level research skills and perform research through the medium of a doctoral dissertation. For some, earning a terminal degree has been a lifelong dream, while others want to use the degree to move into a teaching career. These students are ambitious, highly focused professionals with years of experience on the job, and they view faculty members as coaches and colleagues rather than authorities.

Baby Boomers and Retirees

Baby boomers and retirees are returning to school in large numbers. The number of students age 40 and over has tripled since the 1970s, and 2 million baby boomers are currently taking classes (Thompson, 2009). Many of these older students seek degrees that will allow them to pursue "encore" careers: second careers in different fields from the ones in which they spent the majority of their working life. Members of the baby boom generation often choose encore careers that are personally fulfilling, such as starting a business, or socially redeeming, such as working for a nonprofit. An estimated 6% to 9% of this age group—between 5.3 and 8.4 million people—is pursuing encore careers (The MetLife Foundation & Civic Ventures, 2008).

Societal Trends Driving Working Learners to Seek Education

In the future, the working learner population will likely grow as more people recognize the need for lifelong learning. Three major trends that are driving the influx of working learners into degree programs are projected to intensify in the future: career and educational mobility, the increasing sophistication of the workplace, and the rise of the "ageless society."

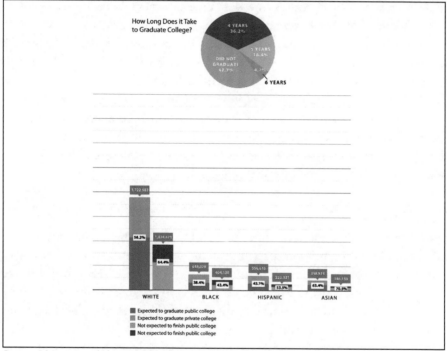

Source: University of Phoenix Knowledge Network. (2010). *Extraordinary commitment: Challenges and achievements of today's working learner*. Phoenix, AZ: Author.

Mobility

In past decades, it was common for workers to remain with the same company for their entire career. That paradigm has shifted: now workers expect to have multiple jobs, and even multiple careers, over their working lifetime. To gain the skills they need for these different careers, workers often return to school for credits or degrees. However, employee mobility provides companies with fewer incentives to make educational investments, even while demand for workforce development increases. For example, 59% of corporate respondents in one survey said they were under pressure to reduce training budgets; at the same time, 24.5% of respondents said firms were placing more emphasis on learning during the economic downturn (Grohs, 2009). As companies make fewer training resources available, employees are turning to colleges and universities to update their skill sets.

A More Educated Workforce

Education begets education. As the workforce becomes more educated, employees need more degrees to stay competitive. In 1950, for example, only 50% of American adults had completed high school, and just 5% held a bachelor's degree or higher (Stoops, 2004). By 2007, 84% of American adults held high school diplomas, and 27% had completed a bachelor's degree or higher (Crissey, 2009). This dramatic upsurge in education means that the bachelor's degree is no longer the mark of distinction it once was; in some circles, the master's degree has become known as "the new bachelor's." Today's careers have also become increasingly specialized and reliant upon technological skills, putting educated workers at a distinct advantage.

The Ageless Society

Americans are living longer—and, just as important, staying healthy longer. Many are choosing to work past traditional retirement age, whether out of necessity, love of their career, or desire for personal fulfillment. For the first time, four generations, ranging from age 18 to over 70, coexist in the workplace: matures, baby boomers, Generation Xers, and Millennials. The longer an employee remains in the workforce, the more likely he or she will need to learn new skills to keep up with business and technological developments. Mid-, late- and second-career workers are therefore seeking degrees and credits to ensure they remain competitive and knowledgeable.

How Higher Education Can Better Meet the Needs of Working Learners

Higher education has become increasingly aware of working learners over the past decade, but most colleges and universities are still centered on the needs of the traditional college student. For instance, their curricula are designed to produce "well-rounded" graduates conversant in a variety of subjects, rather than individuals with specific, job-oriented skills. These curricula often mandate core requirements and electives more appropriate for young people who have recently left high school, as opposed to adults who have been in the working world for years. Classes are typically scheduled during the daytime on weekdays, with few night, weekend, or online options, and many important offices (such as the registrar and the financial aid office) are only open during working hours as well. Long semesters lock students into classes that last months, causing them to lose time, money, and effort if work or life circumstances force them to drop a class. Faculty and support staff often assume that nontraditional students are more knowledgeable about procedures (e.g., choosing and enrolling in classes, navigating the financial aid landscape) and how to "do school" (e.g., study, manage time, use the library, seek help, write long papers, communicate with professors) than they truly are.

Some colleges and universities have made changes to their curriculum, course delivery methods, and support services to better serve working learners. The innovative ways they have done so will be described in later chapters, but the following three principles can serve as a précis for educators interested in better accommodating working students.

Flexibility in Scheduling and Access

The most pressing problem faced by working learners of all educational and income levels is time management. Work and family demands make it difficult for adult students to attend classes held during weekday mornings or afternoons. Flexible class scheduling, including holding classes at night and on weekends, and online classes that allow students to participate at any time, make it possible for working learners to earn degrees despite their many obligations.

Support for Returning Learners

Many working learners have been out of school for years, even decades, and returning to the academic world can cause them considerable anxiety. Some, especially first-generation students, may doubt their ability to succeed in the college classroom, and others may worry that their academic skills have been weakened by long periods away from school. For such students, positive reinforcement may mean the difference between remaining in a program and dropping out. The human touch that faculty and support staff provide is even more critical for working learners than it is for traditional students. Support staff such as academic and financial aid advisors should be trained to respond to the needs of working learners. Such training may include extra instruction about campus procedures for first-generation students and those who have been out of school for a long time. Frequent communication with available, approachable faculty members can also make working learners feel more welcome in the classroom. Many working learners will also need support in the form of remedial or refresher classes; some, particularly first-generation or first-time students, may need help with skills such as studying and time management.

Ensuring That Curricula Are Relevant to Working Learners

As working professionals, adult students want their courses to be immediately relevant to the workplace. Curricula for working learners should consist of rigorous, focused classes that teach a battery of relevant job skills. Course material should be regularly updated to reflect changes in technology and the workplace. Working students also appreciate learning from practitioner faculty: instructors with job experience in the subject they teach, who can help them see the connection between classroom learning and what they do on the job.

The Importance of Lifelong Learning

Despite the fivefold increase in college attainment since 1940, America still needs more educated workers if it is to remain globally competitive. By some estimates, the United States will need 64 million degree-holding workers by 2025 to retain its global prominence. If current completion rates remain constant, the nation will still face a shortage of 16 million degree-equipped workers by that date, which will eventually cause America to lag economically (The National Center for Higher Education Management Systems & Jobs for the Future, 2007). To prevent this shortage, more working adults who are past traditional college age must earn degrees. If they are to do so, however, colleges and universities must reimagine their curricula, scheduling, and services to meet the needs of a broader college-going population.

Summary

Seventy-three percent of students enrolled in higher education today can be classified as nontraditional students or working learners: students over the age of 23 who work part- or full-time while attending school. Many working learners have dependents and are financially independent, and almost all live off-campus. Workplace and societal trends such as worker mobility, the increasing sophistication of the workplace, and the extension of the working lifespan have produced a rise in the working-learner population. This diverse group of students comprises single mothers, first-generation and reentry students, career changers, senior citizens, retirees, and those pursuing advanced degrees.

Working learners have different needs than traditional, 18-to-22-year-old college students. Almost all working learners struggle with time management, and certain subgroups of working learners may need remedial classes, training in academic skills such as studying and performing research, and emotional support from faculty and support staff. To better meet their needs, institutions of higher education should provide flexible class scheduling, expanded support services that account for working learners' unique motivations and requirements, and classes that teach students skills they can immediately put to use in the workplace

References

American Council on Education. (2008, November). *Serving those who serve: Higher education and America's veterans.* Retrieved from http://www.acenet.edu/Content/NavigationMenu/ProgramsServices/MilitaryPrograms/serving/Veterans_Issue_Brief_1108.pdf

Capriccioso, R. (2006, January 26). Aiding first-generation students. Retrieved from Inside Higher Ed website: http://www.insidehighered.com/news/2006/01/26/freshmen

Choy, S. (2002). *Findings from the condition of education 2002. Nontraditional undergraduates* (NCES 2002–012). Retrieved from National Center for Education Statistics website: http://nces.ed.gov/pubs2002/2002012.pdf

Cook, B. J., & Kim, Y. (2009, July). *From soldier to student: Easing the transition of service members on campus.* Retrieved from http://www.acenet.edu/AM/Template.cfm?Section=HENA&Template=/CM/ContentDisplay.cfm&ContentID=33233

Crissey, S. R. (2009, January). *Educational attainment in the United States: 2007.* Retrieved from U.S. Census Bureau website: http://www.census.gov/prod/2009pubs/p20–560.pdf

DiRamio, D., Ackerman, R., & Mitchell, R. L. (2008). From combat to campus: Voices of student-veterans. *NASPA Journal, 45*(1), 73–94.

Grohs, M. (2009). *Training and development.* St. Petersburg, FL: i4cp.

Johnson, J., Rochkind, J., Ott, A. N., & DuPont, S. (2009). *With their whole lives ahead of them: Myths and realities about why so many students fail to finish college.* Retrieved from Public Agenda website: http://www.publicagenda.org/files/pdf/theirwholelivesaheadofthem.pdf

The MetLife Foundation & Civic Ventures. (2008). *Americans seek meaningful work in the second half of life. A MetLife Foundation/Civic Ventures encore career survey.* Retrieved from http://www.civicventures.org/publications/surveys/encore_career_survey/Encore_Survey.pdf

The National Center for Higher Education Management Systems & Jobs for the Future. (2007, November). *Adding it up: State challenges for increasing college access and success.* Retrieved from http://www.ecs.org/html/Document.asp?chouseid=7741

Pascarella, E. T., Pierson, C. T., Wolniak, G. C., & Terenzini, P. T. (2004). First-generation college students: Additional evidence on college experiences and outcomes. *The Journal of Higher Education, 75*(3), 249–284.

Pike, G. R., & Kuh, G. D. (2005). First- and second-generation college students: A comparison of their engagement and intellectual development. *The Journal of Higher Education, 76*(3), 276–300.

Quinn, P., & Allen, K. R. (1989, October). Facing challenges and making compromises: How single mothers endure. *Family Relations,* 390–395.

Stoops, N. (2004, June). *Educational attainment in the United States: 2003. Population characteristics.* Retrieved from U.S. Census Bureau website: http://www.census.gov/prod/2004pubs/p20-550.pdf

Thompson, M. (2009, January 5). More baby boomers return to classroom. Retrieved from LifeWhile website: http://lifewhile.com/money/17505764/detail.html

Tym, C., McMillion, R., Barone, S., & Webster, J. (2004). *First-generation college students: A literature review.* Retrieved from Indiana Pathways to College Network website: http://in pathways.net/first_generation_college_students.pdf

University of Phoenix. (2009). *Student registration survey [of those reporting] and student administration database (OSIRIS).* Accessed October 21, 2009.

U.S. Department of Education, National Center for Education Statistics. (2009). *Digest of Education Statistics, 2008* (NCES 2009–020). Retrieved from U.S. Department of Education National Center for Education Statistics website at: http://nces.ed.gov/pubs2009/2009020.pdf

SECTION II

WORK TRENDS

In this section, we look at the future of women and the future of work, which have led to new economic realities and opportunities for the U.S. Over the years, women have made tremendous strides as key influencers in the U.S. economy by becoming corporate workers, owners of new businesses, and key influencers of family purchasing decisions. Likewise, globalization, small businesses, and workers who are self-employed, freelancers, or who perform microwork continue to bring great changes to the employment landscape. Higher education requirements must cater to these changing economic and work realities and to the rapidly globalizing business world.

· 3 ·

THE FUTURE OF WOMEN

No longer a minority, women now comprise a major force in America. They are significant contributors to the U.S. economy—they influence more than 25% of the U.S. GDP (Barsh & Yee, 2011). Between 1970 and 2009, women went from holding 37% of all jobs to nearly 48%, almost 38 million more women during this period. Without women in the workforce, our economy would be 25% smaller than it is today (Barsh & Yee, 2011, p. 4). According to the U.S. Census Bureau, the number of working women has grown by 44.2% since 1984, up from 44 million to 72 million in 2009 (BLS, 2010).

Women account for over 46% of the workforce (U.S. Department of Labor, 2009). Women aged 25 to 34 are more likely to have a college degree and more likely than men to go to graduate school (Silverman, 2011). Women-owned firms create and/or maintain more than 23 million jobs with an economic impact of $3 trillion annually, which translates into 16 percent of all U.S. jobs (National Association of Women Business Owners, 2009). Corporations such as GM, Shell, McDonald's, Time Warner, and Kaiser recognize that women are the chief consumers for the family and influencers of household products, and they have placed women in key executive roles in their firms to match the face of their primary customers. (2011 Catalyst Awards Dinner, 2011).

Today, women hold a broad range of managerial, corporate, and professional positions that were traditionally held by men. In another important measure, they are surpassing men in educational achievement; for the first time, more women are earning college degrees than men. This fact alone helps position women to prevail in a competitive future. And, according to author Hanna Rosin of *The Atlantic* magazine, history has moved into a new era that signifies the end of men (Rosin, 2007).

Historically, the status of U.S. women has seen steady progress over the past 50 years, due to the passage of several federal regulations. These include the Equal Pay Act in 1963, which requires equal pay for women doing the same work as men; Title VII of the Civil Rights Act of 1964, prohibiting discrimination in the workplace based on race, gender, and ethnicity; and Title IX of the Civil Rights Act, outlawing sex discrimination in education. In 1950 about one in three women participated in the labor force. By 1998, nearly three of every five women of working age were in the labor force. Among women age 16 and over, the labor force participation rate was 33.9 percent in 1950, compared with 59.8 percent in 1998 (BLS, 2000).

Changing Perceptions of Women in the Workforce

Attitudes about women joining the workforce in large numbers have shifted over the decades. The belief that the man should be the breadwinner in the household decreased from 64% overall in 1977 to 41% in 2008, with approximately half of both genders espousing this viewpoint. Among age groups, the biggest shift occurred in those over 63. The percentage of people in that age group who believed men should earn the money dropped from 90% in 1977 to 53% in 2008. A bare majority still believed in the male breadwinner, but the change of attitude was dramatic. In the 43–63 age group in 2008, 41% believed men should earn the money, while 40% of the 29–42 age group and only 35% of those under age 29 believed this (Galinsky, Aumann, & Bond, 2008). It is clear that age is a significant factor in perception of who should be the major breadwinner. However, in another clear trend into the 21st century, there is new awareness: One twentysomething woman reported that her generation just assumes women will work; it is no longer even a question (Shriver, 2009).

A sign of changing attitudes is most apparent in a 2009 survey of 3,413 American adults sponsored by the Rockefeller Foundation and *Time* magazine.

It found that Americans overwhelmingly believed that the increased partici-pation of women in the workforce is a positive thing. Among those 65 or older, 69% saw it as positive. The percentages were much higher among other groups; 85% of those in the 18–29 age group believed it was positive. Among all the adults surveyed, 89% reported they were comfortable with women mak-ing more than their male partners. That included 83% of the male respondents, and 84% of the female ones (Halpin & Teixeira, 2010).

In the 1950s model, a mother who worked outside the home was thought to be neglecting her children and therefore could not be a good mother. Studies show that attitudes about this point have been shifting as well. Between 1977 and 2008, the percentage of men who believed working women could be good mothers rose from 71% to 80%. Women's agreement with this statement increased from 49% in 1977 to 67% in 2008, leaving a third of women who still think it is better for mothers not to work (Galinsky et al., 2008).

Though attitudes have shifted, the prevailing belief of one parent staying at home to care for the children is not feasible for many two-parent families who cannot afford to live on a single income. In these families, women are con-fronted with many challenges, such as balancing parental duties, juggling workplace issues, and moving into careers that have traditionally been the domain of men.

Women's Shifting Roles and Responsibilities

June Cleaver and other ideal housewives of the 1950s did all the shopping, cleaning, and cooking and tended to children's needs. This was part of their job description. Now that no one is staying home all day to attend to these things, the tasks need to be divided up among family members.

Today men no longer carry the burden of being the sole breadwinner in the family. In fact, Boushey and O'Leary (2009) report that men are now more likely to need and want to take time off from work to attend to their family. According to the authors, there are fewer families with a full-time, stay-at-home wife, and they report that men and women must now negotiate the challenges of work–family conflict, such as who will go in to work late to take an elderly fam-ily member to the doctor or stay home with a sick child.

One study found that 56% of men in 2008 thought they did an equal amount or most of the cooking for the family, while only 25% of the women sur-veyed agreed with this perception. A full 70% of women thought they did an equal amount or most of the cooking. The results for cleaning the house were very similar: 53% of men thought they did as much or more cleaning, and only 39% thought their female partners did as much or more. Among the women surveyed, 73% thought they did an equal amount or most of the housecleaning, and only 20% thought their male partners did. Although studies have shown that the partner with the socially sanctioned responsibility has a tendency to believe she is doing more, there is clearly a gap in perception between men and women, and perhaps a gap between male ideals and action (Galinsky et al., 2008).

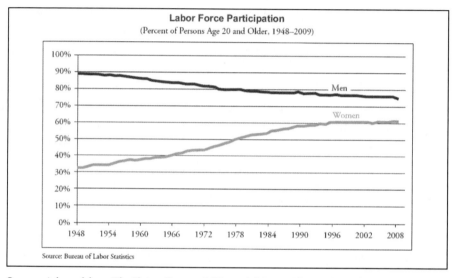

Source: Adapted from *The Shriver Report: A Woman's Nation Changes Everything*, by M. Shriver and The Center for American Progress, 2009, http://www.americanprogress.org/issues/2009/10/pdf/awn/a_womans_nation.pdf

In a September 2009 Rockefeller Foundation and *Time* study, both men and women agreed that women have a disproportionate responsibility for child and elder care, and a majority of men and a slight majority of women thought it better if the woman stayed home and cared for the children. However, there is evidence that spouses and partners are working this out by negotiating with each other instead of relying on preconceived gender roles (Halpin & Teixeira, 2009).

Workplace Attitudes toward Women

In spite of the many compromises men and women make to share parental responsibilities, working mothers have not yet been fully accepted in the workplace. For example, the 2009 Rockefeller/*Time* survey found that 82% of men and 81% of women disagreed with the statement that mothers cannot be as productive at work as fathers. A similar percentage (82% of men and 81% of women) disagreed with the statement that mothers cannot be as productive as people without children (Halpin & Teixeira, 2009). Yet another study conducted in the workplace itself found that employers perceived job candidates who are mothers to be less competent, promotable, or likely to be recommended for management than male candidates or women without children. Mothers were less likely to be hired, and if they were, their salary recommendations were lower. Fathers, on the other hand, were not penalized for parenthood; they were often rated higher than their male colleagues without children (Boushey, 2009).

The extra effort required of women to act as breadwinners, while continuing to perform an unequal amount of housework and childcare and having to deal with outmoded workplace attitudes in the workplace, creates stress. This has been a growing problem over the past 40 years. Changes in male attitudes and behavior are comparatively recent, as reflected in the statistics. In fact, men who take on more responsibility for homemaking and childcare are now reporting increased stress, while stress levels for women are staying fairly stable. The genders also have different stressors; men who are work-centric but also have family and household responsibilities experience more stress, as do men who feel they do not have their supervisor's support for balancing their lives. Work-centered women also experience more stress, but women also report being stressed by the number of hours worked, job pressure, and lack of job satisfaction (Galinsky et al., 2008).

Women's Portrayal in the Media

Attitudes toward women have a big impact on how they are treated in society. As a result, many have expressed concerns about the way the media portrays women. Some complain that to show women police chiefs, surgeons, detectives, lawyers, and lawmakers, without portraying ordinary women getting by in the world, presents a distorted picture of reality and sends the message that things are better than they actually are (Douglas, 2009). Others think portrayals of women in traditionally male roles can provide role models that female students cannot find anywhere else.

Portrayal of women in the media is often uneven and open to debate. Experts and pundits consulted by the media are disproportionately male

(Douglas, 2009). Men are seen more often in media presentations. A study of films made from 1990 to 2006 showed that men appeared 2.71 times more often onscreen than women. On television, women have only 37% to 40% of the roles. Storylines tend to value women characters more for their looks than for their character or behavior, and women are much more likely to appear in provocative outfits (Smith, Kennard, & Granados, 2009). Women are appreciated for glamour, power, and sex (Podesta, 2009).

It is important to note, however, that the media has also provided us with many positive images of successful women. TV talk show host Oprah Winfrey, for instance has often been cited as a powerful role model for women. In a *ForbesWoman* magazine readership survey, Winfrey was considered a "modern successful woman of the times," and appreciated not just for her celebrity status, but also for her business acumen, philanthropic activities, and promotion and mainstreaming of important issues, such as health and finances for women. Winfrey was also named the Most Powerful Celebrity by *Forbes* in 2010, (Casserly, 2010). Other important media figures include Martha Stewart, Ellen DeGeneres, Katie Couric, and Barbara Walters, who are often cited for their pioneering efforts in their genres.

In the news media, women's voices are steadily on the rise. In a 2010 global survey of 130 countries, 24% of those people interviewed, heard, seen, or read about in mainstream broadcast and print news were women, a distinct jump from 1995's survey results, in which only 17% were women (*Global Media Monitoring Project*, 2010). Since the media is a critical vehicle of discourse on many social, economic, and cultural fronts, women's presence and participation is important. The report goes on to say, however, that women continue to be underrepresented as experts and authorities in the global broadcast medium. Less than 20% of all spokespersons are women, and only 16% of all stories have women as their specific focus (*Global Media Monitoring Project*, 2010).

In America, women have made considerable strides in the news reporting field. A study comparing all news programming on the major networks (ABC, CBS, and NBC) during one week in February 2007 with the results of a similar survey from 1987 showed that while men reported 73% of stories in 1987, that number had changed significantly in 2007, with men reporting 48% of stories and women 40% (the remaining 12% were team efforts featuring reporters of each gender; Ryan & Mapaye, 2010). The study also found that in the 20-year period, networks had become more diverse, with women and minorities of both genders anchoring and reporting on a regular basis across the news spectrum.

Women and Occupations: Confronting Stereotypes

Many women continue to enter what have been traditionally women's occupations. These professions often pay less, which is one reason why the earning gap between men and women may exist. However, studies have also shown that these occupations (e.g., nursing, teaching, childcare, home care, dental hygiene, and food preparation and service) have withstood the recent economic downturn better than many traditionally male occupations. Moreover, jobs typically dominated by women are projected to grow the most over the next 10 years. Although these jobs are traditionally considered "women's work" and have always paid less, the "steady accumulation of these jobs," according to one analyst, would add to the economy as a whole (Rosin, 2010).

The flip side, however, is that only 38% of working women in the United States are employed in managerial, professional, and related occupations (Harrington & Lodge, 2009). Men have traditionally dominated the practice of law and medicine, although women now comprise 36.5% of all lawyers and 31.8% of all physicians (Boushey, 2009). With high enrollment rates for women in law and medical schools, these percentages will continue to shift in favor of women. In the legal profession, women seem to be more concentrated in the lower levels, making up 45% of the associates in law firms (Harrington & Lodge, 2009). This may be partly because larger numbers of women have been entering the profession and thus have less experience than their older male colleagues.

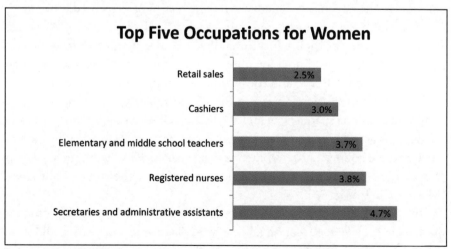

Top Five Occupations for Women

- Retail sales — 2.5%
- Cashiers — 3.0%
- Elementary and middle school teachers — 3.7%
- Registered nurses — 3.8%
- Secretaries and administrative assistants — 4.7%

Source: Adapted from *The Shriver Report: A Woman's Nation Changes Everything*, by M. Shriver and The Center for American Progress, 2009, http://www.americanprogress.org/issues/2009/10/pdf/awn/a_womans_nation.pdf

Women also make up 35.9% of engineers (Boushey, 2009) and 54% of all accountants. Around 30% of accountants are women of color. On the other hand, women make up less than 20% of mechanical and electrical engineers (Harrington & Lodge, 2009), and women are seriously underrepresented as computer software engineers.

Gender Parity in the Workforce

Although women now make up half of the workforce, they are by no means yet an equal half in terms of their income. Despite improvement, gender disparity in earnings still exists. In 1979 the average woman working full time made 62% of what the average man did. By 2007 that percentage had risen to 80%. This is a significant increase in a little less than 30 years, but a 20-point gap remains. Hourly pay for women has increased from 58% of male pay in 1979 to 82% in 2007. This varies by age group. In 2007, women in the 20–24 age group were making 90% of what their male contemporaries made, and those in the 16–19 age group were earning 95%. This difference may have arisen because the workers in the youngest group mostly had unskilled jobs featuring menial work (Galinsky et al., 2008).

A comparison of men and women with similar qualifications who did comparable work shows that women still make only 77% of what their male colleagues make (Taylor, 2010). The pay ratios between women and men also vary somewhat depending on educational levels. Women who did not graduate from high school or who had only a high school diploma in 2008 earned 75% of what men at their educational levels made. Women with some college earned 79.4% of what comparably educated men made, those with college degrees made 78.8%, and those with doctorates or professional degrees made 77.2% (Arons & Roberts, 2009). Gender parity differences can be seen in professions such as law and medicine, where women have made significant progress in employment but are not promoted as often or paid as much as male counterparts. Women lawyers make 77% of what male lawyers earn, while women doctors' salaries are only 59% of their male colleagues' (Mason, 2009).

In spite of these gender-based income differences, however, the progress women have made in education and the workplace in recent years would seem to position them for very real changes and opportunities ahead. Women are major contributors to the overall financial well-being of families and a driving

"Nearly 80 percent of women and men say they are convinced of the benefits of gender parity at all levels. But only about 20 percent believe their companies actually put meaningful resources behind it."
[Harvard Business Review Blog, February 5, 2010]

force of the country's economic growth and prosperity. Writing for *The Atlantic* in an article entitled "The End of Men," Hanna Rosin (2010) alludes to the lingering gender parity in wages but provides an optimistic outlook for the future:

> Yes, the U.S. still has a wage gap, one that can be convincingly explained—at least in part—by discrimination. Yes, women still do most of the child care. And yes, the upper reaches of society are still dominated by men. But given the power of the forces pushing at the economy, this setup feels like the last gasp of a dying age rather than the permanent establishment. Dozens of college women I interviewed for this story assumed that they very well might be the ones working while their husbands stayed at home, either looking for work or minding the children. Guys, one senior remarked to me, "are the new ball and chain." It may be happening slowly and unevenly, but it's unmistakably happening: in the long view, the modern economy is becoming a place where women hold the cards.

Rosin goes on to explain that thinking and communicating, as well as social intelligence and the ability "to sit still and focus," tend to pull women into the open playing field of opportunities and change. These skills are gradually coming to replace physical strength and stamina as the keys to economic success. Countries that rely on the entire population and do not exclude women from participation tend to prosper. These countries are learning that the greater the power of women, the greater the country's economic success (Rosin, 2010). Women are having a dramatic impact on the social and economic makeup of American society, but as with any march to progress, challenges remain that need to be overcome.

Women and Income

Women now bring in 44% of the income in their families and households, and 26% of women earn at least 10% more than the men in those households (Braunschweiger, 2010). Mothers are said to be the primary breadwinners in 40% of households (Shriver, 2009). In the last three decades, the percentage of traditional male breadwinners has dropped from 44.7% to 20.7% (Boushey,

2009). The responsibility increasingly falls upon women to be the primary breadwinners.

In recent years, the earnings of American women has accounted for most of the increases in family income growth. Whereas older generations were able to have a good life on one income, modern families find they need two incomes just to get by (Boushey, 2009). On the other hand, there is evidence that more women are on their own than in any other time in our history. Marriage rates have been dropping for all groups in America since 1970, but the biggest decline has been among less educated women (Taylor, 2010). This means that the women who have to take care of themselves and their children financially are frequently the ones least able to do so.

The percentage of women who bring in more income than their male spouses or partners also differs by economic group. In 2008, 30.1% of women in the top 20% household income group earned as much or more than their male partners, while 67.7% of women in the lowest household income group provided as much or more than their male partners. In the next to lowest quintile, 49.2% of the women brought in as much or more than male partners did. Women in lower socioeconomic groups are thus earning a larger portion of the family income than the income level above them. These figures show a shift over the last four decades for all groups, from 12.6% of the women in the top 20% in 1967, 44% of the women in the lowest 20%, and 28.3% of the next to lowest quintile. In terms of racial groups, 36.9% of White women and 35.8% of Hispanic women have equal or greater income than their male partners, but 51.5% of African American women do (Boushey, 2009).

It is clear, then, that women's contribution in the workforce is essential to the support of the family and household and, by extension, to the structure of our society. The participation of women in the workforce is also essential to the health of the economy, a point that is slowly being recognized by government and business leaders as more women are encouraged to enter high-paying fields such as technology and engineering, where they are underrepresented and their skills are needed.

Closing the Income Gap

Since women comprise almost 50% of the workforce, reaching parity with men's income may be good news for the economy as whole. It is important to note that the gap is closing and for many reasons, including the fact that women only experienced 25% of the layoffs in the 2007–2009 recession (Taylor, 2010).

Another reason for the closure of the income gap may be that the number of women receiving higher education has increased, bringing women into professions traditionally considered male. People who have at least a college education have made greater income gains than those with less education, and this group includes increasing numbers of women (Taylor, 2010).

A third factor behind the narrowing of the income gap may be the fact that men's overall incomes are not increasing as much as women's. Between 1970 and 2007, women's earnings increased by 44%, while the men's earnings rose by only 6% (Taylor 2010). Additionally, male blue-collar workers have been losing their jobs, and other men's wages are actually falling (Kimmel, 2009). Yet even this unequal change has not yet produced parity, and women as a group still do not get equal pay for equal work.

Women have different views on how this income gap should be reconciled. One woman interviewed for *The Shriver Report* in 2009 thought women needed to make more than men, since they have more responsibility for family and childcare. Another woman disagreed, feeling she should get equal pay for equal work and then it would be up to her to decide how to spend the money (Shriver, 2009). Both men and women agree that government and business have not recognized the realities of the modern family and have not done enough to guarantee equal pay so that women can support or help support their families. Moreover, these entities still do not fully understand the need for flexible hours, childcare assistance, and family and medical leave (Boushey & O'Leary, 2009).

Women as Managers and Corporate Leaders

Reaching the ranks of managerial and corporate leadership is on the rise for women, but the road to leadership positions and female representation among management is uneven across all industries. A U.S. Government Accountability Office (2010) report, for instance, showed that in 2009, 40% of managers in the workforce were women, but they had made considerable gains in only 3 of 13 industries analyzed. Women outnumbered or equaled men in managerial positions in healthcare, social assistance (70% of managers were women), and education (57%).

In another area, however, women were making faster progress. Women managers' rate of education achievement has been higher than that of their male counterparts, both in bachelor's degrees and in master's degrees. For instance, the number of men in managerial positions who obtained master's degrees between 2000 and 2007 rose only 1 percentage point, but the number of female managers who held a master's degree or higher rose nearly 4 percentage points. This upward trend in educational achievement is consistent with the national average where women are seeking higher education opportunities at a faster pace than men (U.S. Government Accountability Office, 2009).

In the private sector, the ascent of women to leadership roles has also seen progress but at a slower pace. Women are CEOs at 26 of the Fortune 1000 companies (12 in Fortune 500, 14 in Fortune 1000; Catalyst.org, 2011). Among the 12 in the Fortune 500 are Patricia Woertz from the Archer Daniel Midland Company (ADM), Angela Braly from Wellpoint, Inc., and Indra Nooyi from PepsiCo, Inc.

While these gains may be incremental, there is positive evidence of the influence that women have in these leadership roles. Studies have shown that companies led by women tend to be more financially successful than comparable companies run by men. Fortune 500 companies with women CEOs or more women on their boards had healthier bottom lines than companies where the executives and board members were all men.

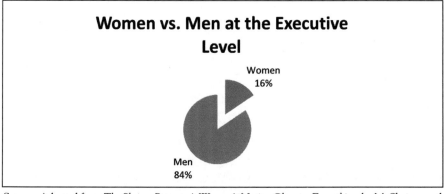

Women vs. Men at the Executive Level

Women 16%

Men 84%

Source: Adapted from *The Shriver Report: A Woman's Nation Changes Everything*, by M. Shriver and The Center for American Progress, 2009, http://www.americanprogress.org/issues/2009/10/pdf/awn/a_womans_nation.pdf

One study of the 25 best corporations for working women showed those companies had 34% higher profits than corporations that were not woman friendly. This may be due to the fact that women make 80% of consumer purchasing decisions and so have a major impact on the economy (Harrington & Lodge, 2009). Women business persons may know better what to market to female purchasers and how to do so.

Career Advancement and Mentorship: Unequal Results?

Traditional corporate cultures function to a large degree through mentoring and networking, and the gender inequity problem is often present here as well. A 2008 online Catalyst survey of more than 4,000 MBA alumni from Canada, America, and Europe showed that a majority agreed that mentors had a high impact on their professional success. However, the same survey also showed that men received greater benefits from mentoring than women. This included higher compensations and more promotions—doing very little to close the gender gap (Carter & Silva, 2010).

One reason for unequal results in mentorship influence may be insufficient numbers of female mentors or contacts who might allow women to advance at the same pace as their male colleagues. Women with caregiving responsibilities cannot participate in many after-hours social activities, and single women may have difficulty managing social relations with male colleagues or may be excluded from couples-only social events (Harrington & Lodge, 2009).

In the same study, Harrington and Lodge discusses issues relating to women's inherent self-perceptions, as well as the manner in which they handle situations and circumstances, which may also impact their career advancement prospects. Women and men tend to communicate differently, which can lead to misunderstandings. Also, women don't always put themselves forward and ask for things, while men tend to ask for what they want. This can mean that women get shortchanged and perceived as not sufficiently aggressive. Women sometimes feel they have to choose between being themselves or playing a role to meet the expectations of male bosses and colleagues.

In another study, women's long-term careers were seen to be impacted significantly as a result of "off-ramping" or taking time off for childcare or other reasons. The study states that 73% percent of women trying to return to the workforce after a voluntary timeout have trouble finding a job. Those who do return lose 16% of their earning power; over a quarter of the women reported a decrease in their management responsibilities, and 22% had to step down to a lower job title (Hewlett, Sherbin, & Foster, 2010).

At a 2011 *Wall Street Journal* Women in the Economy conference, Vikram Malhotra, chairman of the Americas at McKinsey & Co., was among several panelists addressing the issue of the lack of women in senior-level positions. He states it in simple terms: "The corporate talent pipeline is leaky, and it is blocked." In other words, women enter the workforce in adequate numbers, but they begin to drop off when they are first eligible for management positions.

According to Malhotra, many reasons contribute to the leaky pipeline, including structural barriers such as a lack of female role models; their exclusion from informal networks where connections are made; an absence of sponsorship; the 24-hour executive lifestyle and its requirements, which prevent many women from making the leap into high-profile roles; and the individual mind-set itself, in which women's desire to move to the next level dissipates as they age faster than men's ("Where Are All the Senior-Level Women?," 2011).

Researchers have suggested solutions to these barriers, starting with making a "business case" for change that gets people to think differently. This would include strong leadership from the top, refining the organizational process to better track careers and to establish metrics of progress, and encouraging both men and women to become effective sponsors (Barsh & Yee, 2011).

On the other hand, there is a general consensus between men and women that because men are becoming more financially dependent on women, unequal pay and opportunities for advancement affect more than just the women. Therefore, issues such as flexible work hours, work/life balance, childcare, and equal pay are now perceived as being "human" issues that affect the whole of our society and economy. Living a full life, including spending time with family, has become a priority for both women and men, and as both genders ask for flextime, they may be considered insufficiently ambitious or committed to their work (Harrington & Lodge, 2009).

Changing Old World Cultures

Studies have found that the pressures of work are taking their toll in the work/life balance equation. In 1992, 80% of men and 72% of women under 29 reported that they wanted jobs with more responsibility. In 2008, however, there was a distinct drop, as only 67% of men and 66% of women under 29 sought more responsible jobs. A significant drop in this desire for both sexes seems to reflect concerns about job pressures and a perceived lack of flexibility in jobs with more responsibility (Galinsky et al., 2008).

The pressures of work impact women in unique ways as compared to men, which produces different obstacles. Women who have more family-care responsibilities, for instance, are more likely to look for part-time work. When they do, they may come up against resistance from employers to allow employees to work part-time. This resistance results in a significant talent drain that the workplace cannot afford (Harrington & Lodge, 2009).

Women workers are also coming into organizational cultures that were created by and evolved under the influence of men. The "right" way of doing things in these cultures is often a male way of doing things, and women's way of communicating and organizing is not always valued. It is for this reason, researchers believe, that women may do better in industries without such entrenched cultures, like the newer technology industry, created in part by young entrepreneurs (Harrington & Lodge, 2009). Despite this expectation, however, women have not been drawn to technology industry jobs. Although the number of information technology (IT) jobs is increasing, fewer rather than more women have responded to this need by entering the field. In 2006 only 26% of IT professionals were women, and men outnumbered women 6 to 1 in IT managerial and leadership positions (McKinney, Wilson, Brooks, O'Leary-Kelly, & Hardgrave, 2008). By 2010 the picture had improved somewhat, but women in IT still face challenges (Collett, 2010).

One study (Wilen-Daugenti, 2000) indicated that there are many areas of subtle discrimination and assumptions about the relative abilities of the different genders—even in the relatively new IT industry, which is not based in the older corporate culture. This contributes to women's difficulties in the IT field. Another study (Harris, Kruck, Cushman, & Anderson, 2009) argues that women's avoidance of work in technology is partly based on attitudes about women's abilities that start in high school or before, where teachers assume that boys are into machines but girls are not. Universities need to recruit women into IT educational paths, but attitudes that influence girls' early education need to change as well. Harris et al. also argue that the IT industry has a bad reputation as being male oriented and exhibiting gender bias.

Even when women start their own IT businesses they tend to specialize in sectors associated with women, such as management consulting, design, and software development. Women-owned IT firms are smaller than those owned by men, both in number of employees and sales. This may be because women have been discriminated against in IT-management hiring and so lack management experience and expertise. Therefore, although women's business ownership in IT is increasing, women are still at a disadvantage (Mayer, 2008).

Solutions to Work/Life Balance and Success

Men and women generally agree that government and businesses need to do more to promote policies and offer resources that allow families to be more productive and successful at work while maintaining healthy and happy personal lives. Women are finding their own solutions, including starting entrepreneurial ventures, getting college degrees, finding good mentors, and identifying fulfilling careers that provide better wages and support structures for balancing work and family.

Self-Employment among Women

One of the ways women are dealing with challenges in the workplace, such as lack of flexible work time and male-dominated corporate culture, is to start their own business and work for themselves. Between 1979 and 2003 the number of self-employed women doubled. Women now represent 35% of self-employed workers, and they continue to open more businesses than men. The number of women business owners is increasing at the rate of 23%, which is 2.5 times faster than the growth of businesses in general (Shriver, 2009).

Women in Business

Women represent more than one third of all people involved in entrepreneurial activity.

(Source: Global Entrepreneurship Monitor (GEM) 2005 Report on Women and Entrepreneurship)

Between 1997 and 2002, women-owned firms grew by 19.8% while U.S. firms overall grew by 7%

(Source: SBA, Office of Advocacy)

Women-owned firms accounted for 6.5% of total employment in U.S. firms in 2002 and 4.2% of total receipts.

(Source: SBA, Office of Advocacy)

The number of women-owned firms continues to grow at twice the rate of all U.S. firms (23% vs. 9%). There are an estimated 10 million privately held women-owned U.S. businesses. The greatest challenge for women-owned firms is access to capital, credit and equity. Women start businesses on both lifestyle and financial reasons. Many run businesses from home to keep overhead low.

(Source: SBA, Office of Advocacy, and Business Times, April 2005)

Women are more likely to seek business advice—69 percent women vs. 47 percent men.

(Source: American Express)

Retrieved from http://www.score.org/small_biz_stats.html, April 26, 2011.

Though these statistics are encouraging, there are other numbers that speak to the challenges that women often face. Although women run around 10 million businesses with a combined annual sales of $1.1 trillion, a significant number of these businesses are only marginally successful. A study found that 46% of women-owned businesses made $10,000 or less per year, and 80% made less than $50,000. Only 3% of these businesses made $1 million or more, compared to 6% of the businesses owned by men (Harrington & Lodge, 2009).

Although female-owned businesses may not have caught up with those owned by men, women are increasingly choosing entrepreneurship as a viable solution to workplace difficulties. Based on the statistics for companies where women play a major role, we can expect them to realize increased success in the marketplace.

Access to Higher Education

Men have traditionally received more education than women, and this remains a reality in most of the world today (Maslak, 2005). In the United States, however, women are catching up by obtaining more education than men. In the 2005–2006 school year, women received 58% of the bachelor's degrees and 60% of the master's degrees in the U.S. By 2012, women are expected to get 60% of bachelor's degrees, 63% of master's degrees, and 54% of doctoral and professional degrees. The predominance of women students has created some concern among educators who feel a majority of women students will create a gender imbalance. Some schools are being investigated, however, based on the suspicion that top private schools are 6.5% to 9% more likely to admit male students than female students (Rosin, 2010).

For women to comprise the majority of students is a major change from the 1970–71 school year, when men earned 94% of all higher education degrees in America. Traditionally, minority women and women with children in the U.S. have achieved lower educational levels (Taylor, 2010). Today, however, women in all ethnic, racial, and socioeconomic groups are racing past men in obtaining higher degrees. Many of these women are nontraditional students who start their education in community colleges, which provide flexible education for an associate's degree or the first two years of a bachelor's degree (Mason, 2009).

Approximately 95% of community colleges have open admission policies and take all students who meet the basic requirements. Most community college students (61%) are independent of their parents, while most students in four-year colleges (66%) receive support from their parents. Two thirds of these independent students are women. Minority students represent 39% of the student body at community colleges but only 24% at four-year institutions (Mason, 2009).

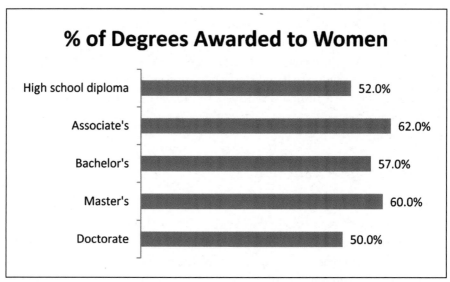

% of Degrees Awarded to Women

High school diploma	52.0%
Associate's	62.0%
Bachelor's	57.0%
Master's	60.0%
Doctorate	50.0%

Source: Adapted from *The Shriver Report: A Woman's Nation Changes Everything*, by M. Shriver and The Center for American Progress, 2009, http://www.americanprogress.org/issues/2009/10/pdf/awn/a_womans_nation.pdf

Choosing Career Paths

Though women are now earning the majority of postsecondary degrees, they have been predominantly for careers in traditionally female areas of work. For example, in 2006 women received 86% of the bachelor's degrees in health professions, including nursing; 78% of the bachelor's degrees in psychology; and 62% of the degrees in biology and medical sciences. They received 50% of business degrees, traditionally a male course of study, but still lagged behind in other areas, earning only 22% of the computer science degrees and 20% of the degrees in some areas of engineering (Mason, 2009).

In 2006, 80.2% of female students pursued traditionally female majors, while men represented 73.1% of computer and information science majors and 92.3% of all majors relating to manufacturing, construction, or repair (Mason,

2009). This situation concerns some commentators, since the traditionally male jobs in computer technology and the physical sciences pay better than many of the jobs for which women are training.

The failure of female students to choose physical science majors may start very young, originating with grade-school teachers who continue to believe that boys are smarter in science and mathematics than girls (Eastin, 2009). Over the past 15 years, there has been some increase in female students in majors where they are typically underrepresented, but this has occurred only after focused effort by government agencies to attract women. For example, NASA has set up an annual Conference for the Undergraduate Women in Physics that attracted over 350 women students in 2009. Top women scientists are featured in the program to provide role models for attendees (Mason, 2009).

Cultures that do not fully use the talents of women are the poorest in the world (Eastin, 2009). We can learn from this mistake by ensuring that we call upon the resources of women's talent in all areas of the job market.

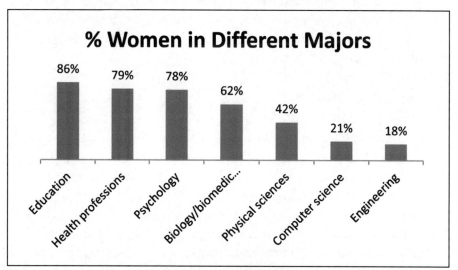

Source: Adapted from *The Shriver Report: A Woman's Nation Changes Everything*, by M. Shriver and The Center for American Progress, 2009, http://www.americanprogress .org/issues/2009/10/pdf/awn/a_womans_nation.pdf

Mentorship and Sponsorship

For most women, role models emerge in everyday life: teachers and coaches who urge girls to reach their potential and individuals who exemplify success through their own achievements. The importance of role models, mentoring, and sponsorship programs is being increasingly recognized and promoted in the business and in corporate worlds as well.

A recent *Wall Street Journal* Women in the Economy Task Force meeting featured prominent women thought leaders, including Justice Sandra Day O'Connor, the first woman to serve on the U.S. Supreme Court; Indra Nooyi (PepsiCo); Sallie Krawcheck (Bank of America); and Marissa Mayer (Google). Justice O'Connor reminded participants of her own quest to be recognized in a male-dominated profession, when as a fresh law student who had graduated at the top of her Stanford University Law School class, no law firm would hire her because she was a woman ("The Global View," 2011). The conference highlighted the work of successful women leaders while charting out important recommendations for the future. Participants noted that, in addition to increasing opportunities for women and fostering environments that promote potential and talent, a key element to success was mentorship: "It's not only about women mentoring women. What's important is for men to mentor and sponsor women as well" (WSJ).

Summary

Women have made significant strides over the past 50 years, breaking through the traditional stereotypes of stay-at-home mom and sole caregiver for children. Perceptions of women in the workforce, and their roles and responsibilities, have evolved dramatically. Women are a vital force in the U.S. economy and have become major participants in higher education programs. While women predominate in professions such as healthcare, finance, retail, and education, they need to make more aggressive strides in higher paying professions, such as science, technology, and engineering fields, and they need to enter management, leadership and executive positions in greater numbers. Likewise, higher education programs must be more cognizant of women as breadwinners and family caregivers and adapt more quickly to their changing role in society.

References

Arons, J., & Roberts, D. (2009). Sick and tired: Working women and their health. In M. Shriver & The Center for American Progress, *The Shriver report: A woman's nation changes everything* (H. Boushey & A. O'Leary, Eds., pp. 123–115). Retrieved from Center for American Progress website: http://www.shriverreport.com/awn/health.php

Barsh, J., & Yee, L. (2011). *Unlocking the full potential of women in the U.S. economy.* (Special Report of The Wall Street Journal Executive Task Force for Women in the Economy 2011) Retrieved from http://online.wsj.com/public/resources/documents/WSJExecutiveSummary.pdf

Boushey, H., (2009). The new breadwinners: Women now account for half of all jobs, with sweeping consequences for our nation's economy, society, and future prosperity. In M. Shriver & The Center for American Progress, *The Shriver report: A woman's nation changes everything* (H. Boushey & A. O'Leary, Eds., pp. 31–69). Retrieved from Center for American Progress website: http://www.shriverreport.com/awn/economy.php

Boushey, H., & O'Leary, A. (2009). Executive summary. In M. Shriver & The Center for American Progress, *The Shriver report: A woman's nation changes everything* (H. Boushey & A. O'Leary, Eds., pp. 17–27). Retrieved from Center for American Progress website: http://www.shriverreport.com/awn/execSum.php

Braunschweiger, J. (2010, September). Attack of the woman-dominated workplace. Retrieved from http://www.more.com/reinvention-money/careers/attack-woman-dominated-workplace

Carter, N. M., & Silva, C. (2010). *Mentoring: Necessary but insufficient for final advancement.* Retrieved from Catalyst website: http://www.catalyst.org/file/415/mentoring_necessary_but_insufficient_for_advancement_final_120610.pdf

Casserly, M. (2010, July 17). The world's most inspiring women. Retrieved from Forbes website: http://www.forbes.com/2010/07/17/role-model-oprah-winfrey-angelina-michelle-obama-forbes-woman-power-women-jk-rowling.html

Collett, S. (2010, August 9). Women in IT: The long climb to the top. *Computer World.* Retrieved from http://www.computerworld.com/s/article/350061/Women_in_IT_The_long_climb_to_the_top

Douglas, S. J. (2009). Where have you gone, Roseanne Barr?: The media rarely portray women as they really are, as everyday breadwinners and caregivers. In M. Shriver & The Center for American Progress, *The Shriver report: A woman's nation changes everything* (H. Boushey & A. O'Leary, Eds., pp. 281–309). Retrieved from Center for American Progress website: http://www.shriverreport.com/awn/media.php

Eastin, D. (2009). Must Jill come tumbling after? In M. Shriver & The Center for American Progress, *The Shriver report: A woman's nation changes everything* (H. Boushey & A. O'Leary, Eds., pp. 194–195). Retrieved from Center for American Progress website: http://www.shriverreport.com/awn/eastin.php

Galinsky, E., Aumann, K., & Bond, J. T. (2008). *2008 National Study of the Changing Workforce. Times are changing: Gender and generation at work and home.* Retrieved from Families and Work Institute website: http://www.familiesandwork.org/site/research/reports/Times_Are_Changing.pdf

Global Media Monitoring Project 2010: Highlights of preliminary findings. (2010). Retrieved from http://www.whomakesthenews.org/images/stories/website/gmmp_reports/2010/gmmp_2010_prelim_key_en.pdf

The global view: Sandra Day O'Connor, Wei Sun Christianson and Robert Zoellick offer a report card on women's progress. (2011). Retrieved from *The Wall Street Journal* website: http://online.wsj.com/article/SB10001424052748704013604576246292633371136.html

Halpin, J., & Teixeira, R. (2009). Battle of the sexes gives way to negotiations: Americans welcome women workers, want new deal to support how we now work and live today. In M. Shriver & The Center for American Progress, *The Shriver report: A woman's nation changes everything* (H. Boushey & A. O'Leary, Eds., pp. 395–417). Retrieved from Center for American Progress website: http://www.shriverreport.com/awn/americanPeople.php

Harrington, B., & Lodge, J. J. (2009). Got talent? It isn't hard to find: Recognizing and rewarding the value women create in the workplace. In M. Shriver & The Center for American Progress, *The Shriver report: A woman's nation changes everything* (H. Boushey & A. O'Leary, Eds., pp. 199–231). Retrieved from Center for American Progress website: http://www.shriverreport.com/awn/business.php

Harris, N., Kruck, S. E., Cushman, P., & Anderson, R. D. (2009, Winter). Technology majors: Why are women absent? *Journal of Computer Information Systems.* Retrieved from International Association for Computer Information Systems website: http://www.iacis.org/jcis/pdf/Harris_etal_50_2.pdf

Hewlett, S. A., Sherbin, L., & Foster, D. (2010, June). Off-ramps and on-ramps revisited. *Harvard Business Review.* Retrieved from http://hbr.org/2010/06/off-ramps-and-on-ramps-revisited/ar/1

Kimmel, M. (2009). Has a man's world become a woman's nation? In M. Shriver & The Center for American Progress, *The Shriver report: A woman's nation changes everything* (H. Boushey & A. O'Leary, Eds., pp. 323–357). Retrieved from Center for American Progress website: http://www.shriverreport.com/awn/men.php

Maslak, M. A. (2005, October-December). Higher education and women: Deconstructing the rhetoric of the education for all (EFA) policy. *Higher Education in Europe, 30*(3–4), 277–294.

Mason, M. A. (2009). Better educating our new breadwinners: Creating opportunities for all women to succeed in the workforce. In M. Shriver & The Center for American Progress, *The Shriver report: A woman's nation changes everything* (H. Boushey & A. O'Leary, Eds., pp. 161–193). Retrieved from Center for American Progress website: http://www.shriverreport.com/awn/education.php

Mayer, H. (2008, December). Segmentation and segregation patterns of women-owned high-tech firms in four metropolitan regions in the United States. *Regional Studies, 42*(10), 1357–1383.

McKinney, V. R., Wilson, D. D., Brooks, N., O'Leary-Kelly, A., & Hardgrave, B. (2008, February). Women and men in the IT profession: Fewer women entering IT drives the underrepresentation problem. *Communications of the ACM, 51*(2), 81–84.

National Association of Women Business Owners. (2009). *The economic impact of women-owned businesses in the United States.* Retreived from http://www.nwboc.org/media/CFWBR%20report%20Economic%20Impact%20WBO.pdf

Podesta, J. D. (2009). Preface. In M. Shriver & The Center for American Progress, *The Shriver report: A woman's nation changes everything* (H. Boushey & A. O'Leary, Eds., pp. i–v).

Retrieved from Center for American Progress website: http://www.shriverreport. com/awn/preface.php

Rosin, H. (2010, July/August). The end of men. The Atlantic, 56–70.

Ryan, K. M., & Mapaye, J. C. (2010). Beyond anchorman: A comparative analysis of race, gender, and correspondent roles in network news. Electronic News, 4(2), 97–117. doi: 10.1177/1931243110367755

Shriver, M., & The Center for American Progress. (2009). The Shriver report: A woman's nation changes everything (H. Boushey & A. O'Leary, Eds.). Retrieved from Center for American Progress website: http://www.shriverreport.com/awn/index.php

Silverman, R. E. (2011, March 1). The state of American women. Retrieved from The Juggle blog, The Wall Street Journal website http://blogs.wsj.com/juggle/2011/03/01/the-state-of-american-women/

Smith, S. L., Kennard, C., & Granados, A. D. (2009). Sexy socialization: Today's media and the next generation of women. In M. Shriver & The Center for American Progress, The Shriver report: A woman's nation changes everything (H. Boushey & A. O'Leary, Eds., pp. 310–317). Retrieved from Center for American Progress website: http://www.shriverreport. com/awn/socialization.php

Taylor, P. (2010, January). Women, men and the new economics of marriage. Retrieved from Pew Social & Demographic Trends website: http://pewsocialtrends.org/files/2010/11/new-economics-of-marriage.pdf

2011 Catalyst Awards Dinner. (2011). Retrieved from Catalyst website: http://www.catalyst. org/page/70/2011-catalyst-awards-dinner

U.S. Bureau of Labor Statistics. (2010). Women in the labor force: A databook (2010 edition). Retrieved from http://www.bls.gov/cps/wlf-intro-2010.htm

U.S. Department of Labor. (2009). Quick stats on woman workers, 2009. Retrieved from http://www.dol.gov/wb/stats/main.htm

U.S. Government Accountability Office. (2009). Women in management: Female managers' representation, characteristics, and pay. Retrieved from http://www.gao.gov/new.items/ d101064t.pdf

Where are all the senior-level women? (2011, April 1). Retrieved from The Wall Street Journal website: http://online.wsj.com/article/SB10001424052748704013604576246774042116558 .html

Wilen-Daugenti, T. (2000). A study of current attitudes on utilization and satisfaction of women who work for high-tech firms. Unpublished manuscript.

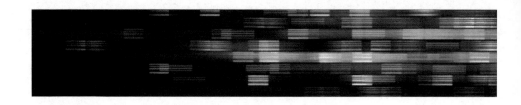

· 4 ·

THE FUTURE OF WORK

In this chapter, we discuss key work trends such as globalization, the rise of virtual organizations and telecommuting, the growth of small businesses and entrepreneurship, and the expansion of freelance ecosystems and microwork. Use of technology as a connector and facilitator of business have inspired these new work environments and transformed the way we conduct business and earn our livelihoods. They are also shaping the skills required for future work and imposing new demands on higher education.

To remain employable today and into the future, workers will need to develop higher-order thinking skills and be well versed in emerging technology. This shift was confirmed by the University of Phoenix Research Institute in a recent study it conducted with the Institute for the Future, a nonprofit strategic forecasting group. Their *Future Work Skills 2020* report analyzes key forces that will redefine the jobs of the future and forecasts the worker skills and knowledge that will be integral to these future jobs (Institute for the Future & The University of Phoenix Research Institute, 2011). For example, cross-cultural competency, or the ability to operate effectively in different cultural settings, was identified as one of many important skills needed for the future. (A closer look at new technologies reshaping work and society is undertaken in Section III. The *Future Work Skills 2020* report is discussed in more detail in Chapter 9.)

Another study (Meyer, 2010a) highlights recruitment and hiring issues manufacturers face today in the ultra-competitive global marketplace. Smart technologies are changing the nature of manual-labor tasks, suggesting the need for industry-wide credentials to ensure that workers can meet the new skill demands of the future.

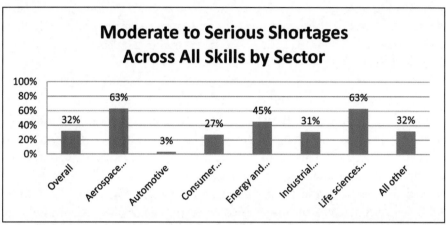

Source: Deloitte, Oracle, and The Manufacturing Institute (2009), *People and Profitability: A Time for Change. A 2009 People Management Practices Survey of the Manufacturing Industry*, Washington, DC: The Manufacturing Institute, http://www.deloitte.com/assets/Dcom-UnitedStates/ Local%20Assets/Documents/us_pip_peoplemanagementreport_100509.pdf. Cited in "Manufacturers Seek Out Qualified Workers," University of Phoenix Knowledge Network, http://www.phoenix.edu/uopx-knowledge-network/articles/industry-viewpoints/manufacturers-seek-out-qualified-workers.html

Holding a single job for one's entire career until retirement and a gold watch has given way to having 10 different jobs and employers over the course of one's lifetime (BLS, 2010). Today, we have virtual organizations and ecosystems. Telecommuting is up 400%, and freelancers (of whom there will be an estimated 14 million by 2015) and microworkers are predicted to become a significant part of the U.S. workforce by the end of this decade. New technologies and social media platforms have opened the way for collective intelligence based on virtual social connections. We are swiftly moving away from assured employment and into a free-market economy for educated, skilled, technically capable talent, and the impact on society and education has been profound.

Growth of Globalization

Technology has changed the face as well as the pace of the American economy. The old ways of doing business within geographical boundaries have given way to exchanging goods and services across countries and continents in an instant, resulting in a diverse global workforce. Much like the Industrial Revolution, which led to mass manufacturing and a new economic framework in the early decades of the 20th century, computer and Internet technology have dramatically altered the American business landscape in the 21st century, in ways both large and small.

Outsourcing, offshoring, and global product development are characteristics of a globalized business (Edmundson, 2009). *Outsourcing* refers to contracting with individuals or businesses outside their home country to perform work, develop a product, or manage some aspect of the business. *Offshoring* means relocating a business, factory, or manufacturing plant to another country. For example, cars and pharmaceuticals may be manufactured in plants run and owned by a U.S. business but located in Southeast Asia, India, or China. *Global product development* involves collaboration among workers from different countries to develop a product.

Experts agree that globalization, defined as "the increasing integration of economies around the world" through the movements of goods and services, and more recently, people and knowledge, is gaining momentum at a fast clip. The question is not whether it will continue but at what pace (International Monetary Fund, 2008). Globalization also means increased competition. According to Meister and Willyerd (2010, p. 21), the speed of globalization is rapidly accelerating. For example, they noted if a firm was on the Fortune 500 list in 1980, there was a 56% chance it was still listed in 1994. However, if a firm was listed in 1994, there was a 30% chance it was still listed on the list in 2007.

To illustrate this trend further, Meister and Willyerd (2010) point to the Financial Times Global 500 rankings showing that between 2005 and 2009, the number of Global 500 companies headquartered in Brazil, Russia, India, and China (the so-called BRIC countries) saw significant growth, with China increasing the total number of headquarters based there by 438%, India 100%, Brazil 80%, and Russia 50%. Global 500 headquarters based in the United States, however, decreased by 17%. One reason may have been that during the recession that began in 2007, growth slowed in developed countries such as America, members of the European Union, and Japan while it gained momentum in developing areas of Asia, such as India and China.

The second reason may be that less expensive labor in developing countries may also constitute a great deal of global business. Labor costs in emerging markets, for example, average 20% to 80% of labor costs in the U.S. in spite of recent increases in wages in developing countries (Hansen, 2006). America is also experiencing labor shortages in key areas, especially in the technology field, due to demographic changes. Many skilled workers are retiring, and there are not enough Generation X workers available to replace them (Klein, 2009).

It is clear that every aspect of a business today often involves multiple collaborations, from concept design and engineering to marketing and customer service. This means that local workers have to compete in a global market to keep their positions. In a larger context, it also means that employees who have education, proper training, and technology skills can be the differentiating factor that makes them indispensable to their company, which in turn becomes more attractive to potential overseas partners. In fact, manufacturers and small business owners are learning today that to compete in an international market, they need to be trained and well qualified in all areas of their business, including entrepreneurial ventures, sales, marketing, information technology, and engineering.

Globalization has also meant that more natives of their employer's home country work overseas. In the United States these sorts of job assignments have increased recently. One 2009 survey showed that almost half of the respondents had stepped up the number of workers sent abroad for 1- to 5-year assignments, with 38% of those surveyed increasing the number of employees who work abroad on multiple assignments. Experts claim this trend is irreversible and that a culturally diverse workforce is here to stay (Klein, 2009).

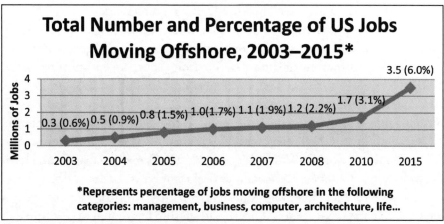

Total Number and Percentage of US Jobs Moving Offshore, 2003–2015*

*Represents percentage of jobs moving offshore in the following categories: management, business, computer, architechture, life...

Source: "Trends in Offshoring," B. Donahue, 2007, *Business Facilities: The Location Advisor*, http://www.nxtbook.com/nxtbooks/groupc/bf_200712/index.php?startid=48

Rapid technological changes over the past few decades have also paved the way for a permanently globalized business environment. While companies are beginning to adjust to this new reality, many struggle with important aspects of managing, training, and nurturing a globalized workforce.

Global Business Challenges

Meister and Willyerd (2010, pp. 23–24) note that the nature of work is rapidly changing in the global marketplace. More than in the past, firms today are tapping into a large pool of worldwide workers. Many firms have created a large global footprint by virtualizing their workplace. Cisco Systems, for instance, established a global development center with over 7,000 employees in Bangalore, India (Cisco India Overview, n.d.), the largest outside of the U.S. IBM opened its procurement offices in Shenzhen, China, in 2006 (Business Wire, 2011), and McDonald's—widely considered to be a forerunner in globalizing—has opened 17,000 of its over 31,500 worldwide restaurants outside of America (Bnet.com, 2005).

Doing business in the global arena raises challenges that many businesspeople from developed nations have never confronted before. Perhaps because of this lack of experience, the failure rate for global efforts has been estimated to be as high as 57% (Edmundson, 2009). There are several possible factors behind this. For one, when companies are offshoring or conducting global market development, they need to start with the right balance of workers from the company's home country, the country where the work is being done, and third-party nations (Tarique & Schuler, 2008). This will help ensure that the work is consistent with the company's quality requirements.

Ineffective leadership can also make doing business globally difficult. Capable leadership is an essential aspect to business success, yet the traits that are effective in one culture may be far from effective in another. What worked for years in the home country may not work abroad (Klein, 2009,). In China, for example, where the culture values collectivism, a leader's skill at managing relationships (*guānxi*), is often the most important leadership trait (Edmundson, 2009). Negotiating styles may also need adjustment to suit Chinese cultural norms (Wilen-Daugenti, 2007). Foreign managers and leaders who do not understand the culture or its emphasis on the collective may not manage relationships well, and consequently will not be perceived as effective or competent leaders. Effective leadership has to function within cultural bounds. For example, financial incentives for workers are often popular in America, but they may be perceived as insulting in some other cultures; when they are acceptable, they have to be in appropriate amounts for the country where the workers are located (Brannen & Wilen-Daugenti, 1993; Klein, 2009; Wilen, 2000a; Wilen-Daugenti & Wilen, 1995; Wilen-Daugenti, 2002, 2007).

A leader who is successful in a culturally diverse setting has qualities and traits not necessarily found among managers in the home country. The international leader has to be patient and willing to assimilate into a new culture. Companies often look for people skilled in multiple languages, experienced with international travel, and equipped with certain personality traits. These include being open-minded, intelligent, and able to relate to and "play well" with other people. Such traits tend to influence how leaders operate in the workplace. For example, effective global managers often use an approach that includes, empowers, supports, and nurtures his or her culturally diverse workforce (Klein, 2009; Wilen, 2000b; Wilen & Wilen, 1995; Wilen-Daugenti, 2002, 2007, 2010). Tactics employing division and competition to spur workers on may be effective in some cultures, but they are not usually effective in a diverse cultural group where some people are from cultures that put a high value on the good of the group.

Successful expatriate workers on foreign assignments demonstrate similar traits, including cultural intelligence and flexibility, a global mindset, and an orientation toward people (Tariquea & Schuler, 2008). Cultural awareness does not necessarily require workers to be experts on the culture of each country. Rather, it requires people who know what to look for and how to find the right answers (Edmundson, 2009).

The Role of Technology in Global Business

According to a Purdue University Center for Advanced Manufacturing (2005) survey of 2,800 manufacturing companies and 11,056 service sector companies, 96% and 92% of these companies, respectively, used the Internet to conduct business. In addition, 67% also had Web pages.

The study noted that technology is a critical factor in overcoming geographic limitations of doing business and helps firms compete globally. Multiple firms are using a variety of technologies to help close the global gap. Many of these technologies are discussed in the technology section of this book. They include Cisco Systems Inc. TelePresence™ and HP's visual collaboration systems, which offer advanced video conferencing technologies.

A Crimson Consulting Group research study describes Cisco's Tele-Presence™ as delivering virtual interactions using advanced visual, audio, and collaboration technologies. These technologies transmit life-size, high-definition images and spatial discrete audio to make it possible for users to communicate as naturally as they would in person. The study discusses how TelePresence™ has a quick return on investment (ROI) or payback (14 months) with firms due to reduced travel costs (Crimson, 2009). Similarly, Web-based conference products such as Cisco's WebEx (webex.com) is used extensively in corporate settings, universities, finance, marketing, manufacturing, human resources, and IT companies to bridge distances and reduce travel costs.

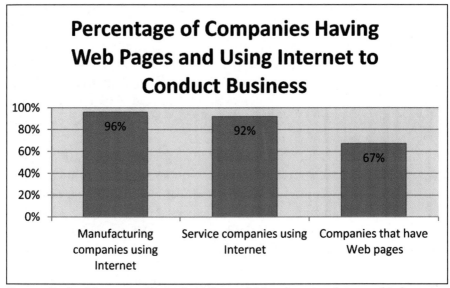

Source: Purdue University Center for Advanced Manufacturing, Impact of Information Technology on Global Business, 2005, Chapter 4, http://globalhub.org/resources/2704/download/003_Impact_of_IT_on_Global_Business_and_Leaders.pdf

Enabling Trust in the 'Virtual' Business World

One of the key ingredients of success in the global business world is trust. Co-workers need to trust each other in order to create goods and services, workers need to trust their managers and business leaders, and customers need to trust businesses. Traditionally, trust in the workplace was based on one-to-one, face-to-face interactions. Workers would meet, talk, and collaborate in the same place. That still happens to a certain degree, particularly in small businesses, but in the ever-growing virtual marketplace, businesses now have to examine how they can create trust in the virtual workplace or business environment, where workers and business partners may rarely, if ever, be in physical proximity to each other.

Researchers have found that trust between co-workers is essential if they are to work successfully on teams. This is even more important if these are virtual teams who don't ordinarily meet (Yakovleva, Reilly, & Werko, 2010). When co-workers trust one another, they will be more willing to do difficult things and take risks, because they feel supported by the rest of the team (Monalisa et al., 2008). A trust-based relationship with co-workers can also create a positive feeling about the company or organization; if the co-workers share this positive outlook, it can affect organizational outcomes (Tan & Lim, 2009). High-tech work environments, computer-mediated communication, and the Internet as a business environment have raised issues of trust among individuals and organizations in academic scholarship as well (Latusek & Gerbasi, 2010).

Trust has more general benefits as well. When people have a high level of trust, they are inclined to give the other the benefit of the doubt and overlook negative information (Earle, 2009). This can allow room for mistakes and downturns without creating a crisis or ending a business relationship. As one group of authors put it, trust "opens the doors to communication, innovation and cooperation, and the retention and achievement of a common goal" (Monalisa et al., 2008).

Commentators recommend a variety of approaches for creating trust in the business context. The first approach is one used to build trust in all human relationships: good communication (Sokjer-Petersen & Thorssell, 2008). Studies have shown that good communication is essential to building and maintaining trust. It can help correct faulty information about another party that might hinder the formation of trust. This communication needs to be effective across language and cultural barriers.

Researchers believe that there is a strong correlation between trust and productivity (Asherman, Bing, & Laroche, 2000). Some believe that it would desirable for team members to develop "identification" or a sense of belonging to groups with shared experiences, to develop cohesion as a group (Fiol & O'Connor, 2005). Since pure virtual environments are often more ambiguous and more uncertain than face-to-face interactions, identifying with groups helps reduce this uncertainty.

Another study (Bjørn & Ngwenyama, 2009) examines the risk of communication breakdowns due to cultural and organizational structures in two globally distributed teams. The authors of this study conclude that "shared meaning," or the capacity to make sense of one another's actions, is an important factor in virtual team collaborations. Shared meaning, the authors add, develops over time and through face-to-face encounters.

Effective collaboration requires trustworthiness across people, processes, and technology (The *Economist* Intelligence Unit, 2008). In a survey of 453 businesses worldwide, an *Economist* survey found that while most collaborators don't wholly trust people and organizations they work with, issues of trust rarely dismantled collaborations. Most forgive lapses in judgment, but they are less forgiving of malicious intent. The survey also found that trust, as well as project success, appeared to decline as collaborations became more and more virtual.

Fifty percent of the people surveyed, however, said although they had "little" trust in an organization with which they had recently collaborated, the collaboration itself was fairly successful, with 6% saying it was *very* successful. The survey also showed that overall, 17% favored face-to-face for more than half their collaborations and 37% used face-to-face for 50% of their collaborations. Also, those who considered themselves to be very good collaborators said 38% of their collaborations were a mix of virtual and face-to-face. Face-to-face meetings were considered important for initial meetings to set goals and make sure everyone was on the same page; once the collaboration was underway, they used technologies such as telepresence (*Economist*, 2008).

In sum, these studies seem to suggest that while virtual teams and electronic communication are an accepted business norm, face-to-face interactions facilitate positive employee relationships in building trust and understanding.

Globalization, Education, and the Future of the U.S. Workforce

America has often been referred to as a melting pot or a blending of races, peoples, or cultures ("Melting Pot," n.d.). Later observers have also compared it to a salad bowl ("Salad Bowl Theory," n.d.), indicating that as a result of multiple waves of immigration, America is now a collection of ingredients, each retaining their own flavor while contributing to the aggregate salad. Regardless of the metaphor, the United States is becoming more racially and ethnically diverse.

Immigration historically has been a major contributor to the U.S. population (Shreshta & Heisler, 2011). The number of foreign-born people now residing in the United States is higher than at any point in U.S. history. At 12.5% of America's 2008 population of 304.1 million, their numbers have reached a proportion not seen since the early 20th century—and they are growing far more rapidly than the native-born.

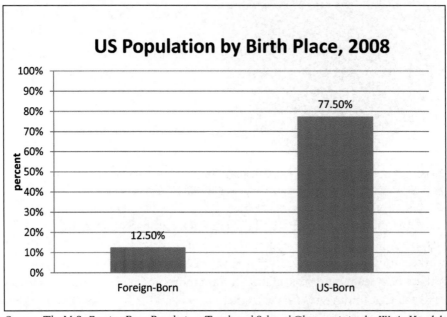

Source: *The U.S. Foreign-Born Population: Trends and Selected Characteristics*, by W. A. Kandel, 2011, retrieved from http://www.fas.org/sgp/crs/misc/R41592.pdf.

According to the Central Intelligence Agency (CIA) World Factbook of April 26, 2011, the U.S. has an estimated population of 313,232,044 (July 2011 estimate), and is 79.96% White, 12.85% Black, 4.43% Asian, 0.97% Alaska Native and Native American, 0.18% native Hawaiian and other Pacific Islanders, and 1.61% of two or more races (July 2007 estimates; CIA, 2011).

The globalized nature of business, coupled with an increasingly diverse U.S. population, has led employers and workers to take a fresh look at the role education will play in the future. In a recent University of Phoenix Research Institute study entitled *The Great Divide* (Heitner & Miller, 2011), researchers found that there were differing views between job seekers and employers about

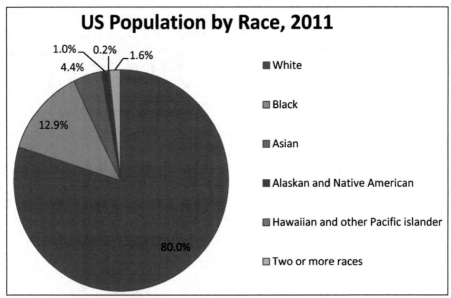

US Population by Race, 2011

1.0% — 0.2% — 1.6%
4.4%
12.9%
80.0%

- ■ White
- ▨ Black
- ■ Asian
- ■ Alaskan and Native American
- ▨ Hawaiian and other Pacific islander
- ▢ Two or more races

Source: Central Intelligence Agency, "United States," *CIA World Factbook*, retrieved from https://www.cia.gov/library/publications/the-world-factbook/geos/us.html.

the demands of the economy and the employment marketplace. In particular, significant disparities were apparent in the area of workers' skills perceptions: workers usually rated themselves highly in being able to work independently, as part of a team, and in a multicultural environment. However, employers reported that finding workers with these types of skills is harder than workers' self-ratings would indicate.

The same study also compared worker and employer of the levels of education necessary for the jobs of the future. Workers rated the demand for bachelor's and master's degrees lower than employers did. Workers also expressed greater uncertainty regarding the demand for higher education in the workplace. To put this issue in perspective, it is important to note that current research suggests that jobs requiring at least a bachelor's degree are growing two times as fast as the overall average (Heitner & Miller, 2011).

In another University of Phoenix Research Institute study (Rouse & Cline, 2011), it was found that the individual ROI of a degree was greater for an employee returning to school (22%) than for an employee just out of school (12%). While both netted a positive ROI, what is clear is that sought-after job skills, combined with education, have a positive impact on the wallet or purse.

Heitner and Miller (2011) also focused on the demand for foreign language proficiency versus the number of workers with this type of skill. The percentage of workers able to conduct business in Spanish, Arabic, Chinese, or Russian is insufficient to meet employers' current demand. In addition, the number of workers who intend to learn a foreign language in the next 10 years will not

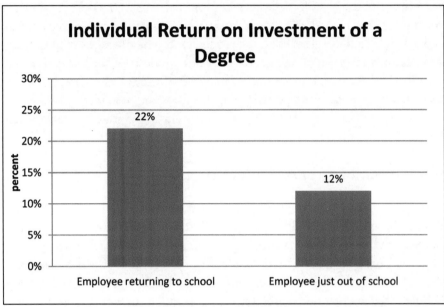

Source: Adapted from *Traditional and Nontraditional Students: Is a Bachelor's Degree Worth the Investment?* by R. A. Rouse and H. M. Cline, 2011, University of Phoenix Research Institute.

be enough to meet the rising demand that employers anticipate.

Growth of Small Businesses and the Independent Worker

Much of the talk about the global marketplace focuses on large corporations, which were among the first to venture into outsourcing, offshore development, and international collaboration in product development. It is important to keep in mind that the U.S economy is largely based in small businesses. Businesses with fewer than 100 employees represented 98% of all U.S. firms with employees in 2004.

Since small businesses fuel innovation and entrepreneurship, many experts consider the state of small businesses a measure of economic health (McArdle, 2010) and believe small businesses are essential to economic recovery in difficult times (Port, 2010). Studies of two recessionary years, 1991 and 2001, show that the smallest firms, including those that start out as one self-employed person, create the most jobs during recessions (SBA, 2009). By employing half the private-sector workforce and creating new jobs, small businesses provide half the nation's nonfarm gross domestic product and generate many of its business innovations. They sometimes inspire whole new industries, further spurring job creation and economic expansion (SBA, 2009). Indeed, the same traits that make small businesses so important to the health of the economy and to its post-recession recovery—innovation, flexibility, dynamic growth—make them viable players in the international business world as well.

Startups and Entrepreneurial Activities

A startup company, or "startup," is a business with a limited operating history ("Startup," n.d.). Generally newly created, startups spend some of their early months and years in a phase of development, during which they research potential markets, perfect their product or service, and in some cases seek investors' capital. The term became internationally popular during the dot-com bubble, when a great number of dot-com companies, particularly in the high-tech field, were founded.

Many entrepreneurial ventures today are a product of recent technological advances. Technology startups fueled the growth of Silicon Valley over the past four decades, and they continue to remain a vital benchmark of economic prosperity. Many tech startups are launched by employees of larger companies who decide they want to strike out in new directions to put their ideas and concepts to work.

A study of U.S. entrepreneurs shows that they come from diverse socio-economic environments. A 2009 Kauffman Foundation study of over 500 companies shed light on the backgrounds, motivation, and beliefs of many of America's entrepreneurs (Wadhwa, Aggarwal, Holly, & Salkever, 2009). The median age of company founders, they learned was 40. Ninety-five percent had earned bachelor's degrees, and 47% had more advanced degrees. Close to 75% of entrepreneurs came from various middle-class backgrounds, and more than half were the first in their families to launch a new venture. A majority had opened their companies after working for another company for more than six years. Less than 1% came from extremely rich or extremely poor backgrounds.

The essential ingredient of the entrepreneurial mindset is the willingness to take on calculated risks—be they in time, financial investments, or career. They have the capacity to form effective venture teams and can identify crit-

An extensive report released in November 2008 by the U.S. Small Business Administration found that small firms had a higher percentage of patents per employee than larger firms and that younger firms were more likely to have a higher percentage of patents per employee than older firms. (Kauffman Foundation, 2009)

ical resources and fundamental skills for building a solid business plan. Last, entrepreneurs possess the vision to recognize opportunity where others see chaos, contradiction, and confusion (Kuratko, 2007).

While successful businesses and ventures receive much attention and praise in the media, close to half of new firms don't last beyond five years (Kauffman, 2010a). Nevertheless, the sheer number of new firms established each year make entrepreneurial ventures in the U.S, an enduring activity. Ninety-two percent

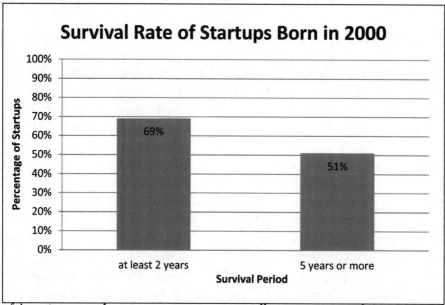

of Americans say that entrepreneurs are critically important to job creation, and more than three quarters of entrepreneurs surveyed believed that economic

recovery is not possible without entrepreneurial activity.
Source: U.S. Small Business Association, "What Is the Survival Rate for New Firms?" retrieved
from http://www.sba.gov/advocacy/7495/8430.

The recession that began in December 2007 is a study in the strength of the
entrepreneurial spirit in America. Small businesses were hit hard during this
period, and many of them had difficulty finding financing to grow or to just keep
afloat. A large percentage has had their applications for financing denied
(McArdle, 2010). Nevertheless, the SBA notes that small businesses will have
an important role in bringing about economic recovery because they are flexible, creative, and innovative, which fosters creation of new jobs in new industries (SBA, 2009).

Another study (Kauffman, 2010b) found that challenging economic times
were a motivating factor for the creation of business. It notes that in 2009,
entrepreneurial activity reached a 14-year peak, exceeding even the boom of
the 1999–2000 dot-com and technology boom. African Americans and older
Americans saw the greatest increase in business creation during this period. This
suggests that the entrepreneurial spirit is hard to suppress even during recessionary times as Americans try to find ways to cope and earn a living.

Women-Owned Businesses

Experts believe that women entrepreneurs may have the potential to jumpstart
the economy during recessions. There are an estimated 8 million majority-
owned (51% or higher), privately held businesses owned by women in the
United States. Women-owned businesses contribute nearly $3 trillion to the
national economy and create or maintain 23 million jobs. These companies
generate 16% of employment in the country (Center for Women's Business
Research, 2009). The businesses women choose to start and run are very
diverse, ranging from professional, scientific, and technical services to administrative support, healthcare, and business services.

Women entrepreneurs cite flexibility as a primary incentive to start their
own business. However, while planning to launch these new ventures they
often find it difficult to identify potential funding sources, lenders, and investors.
Other obstacles that women face are a lack of business-finance knowledge,
inexperience with long-term financial planning, keeping current with new
technology, and the legal issues involved with hiring and managing employees
(Meyer, 2010b).

Another challenge that past women entrepreneurs faced is that many have not attended college or pursued advanced degrees. A 2003–2006 Small Business Office of Advocacy report showed that 28% of women entrepreneurs had a high school diploma, and only 33% had attended some college. However, the report also noted that women entrepreneurs who set educational objectives were seen as more likely to complete their education at higher rates than men.

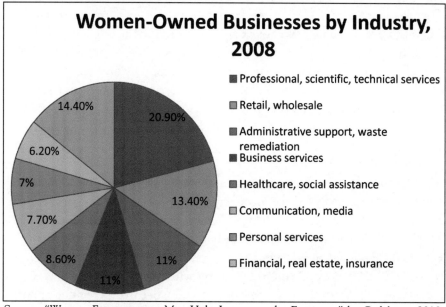

Women-Owned Businesses by Industry, 2008

- Professional, scientific, technical services
- Retail, wholesale
- Administrative support, waste remediation
- Business services
- Healthcare, social assistance
- Communication, media
- Personal services
- Financial, real estate, insurance

20.90% · 14.40% · 6.20% · 7% · 7.70% · 8.60% · 11% · 11% · 13.40%

Source: "Women Entrepreneurs May Help Jumpstart the Economy," by C. Meyer, 2010, www.phoenix.edu/uopx-knowledge-network/articles/industry-viewpoints/women-entrepreneurs-help-jumpstart-economy.html.

The SBA report highlighted a key issue among women entrepreneurs: Because they stand to benefit significantly from education as a way to improve their skills and turn their businesses into more profitable ventures, they can positively impact the economy as a whole (University of Phoenix, 2010).

Today, women in the U.S. earn more advanced degrees than men do, according to Census data, contributing to a historic shift in gender roles. About 10.6 million American women have at least a master's degree, compared with 10.5 million men, but women still lag in fields such as business, science, and engineering. In addition, women are more likely than men to pursue advanced degrees. It is clear that women are in a position to accelerate the growth of the U.S. economy, ensure continued employment, and carry the country forward.

Self-Employment, Freelancing, and Microwork

Self-employed people traditionally have tended to be White, male, married, and older, but that picture, like most of those in the business world, is changing. Though older White males still predominate, numbers of self-employed women and minorities have increased over the last decade. Hispanics have made particularly strong gains in this area, with the number of self-employed Hispanics in the United States doubling since the year 2000; they now comprise 10.3% of the self-employed, up from 5.6%. Immigrants in general have been starting their own businesses at a faster rate than native-born Americans. They now make up 16.5% of the self-employed, up from 12.7% in 2000. A study in 2008 showed that immigrant entrepreneurs produced nearly 12% of U.S. business income (SBA, 2009).

Age and education are also important demographic factors in self-employment. Workers under 35 make up only 15% of the self-employed, and the percentage of self-employed people age 55–64 increased from 16.4% in 2000 to 21.9% in 2007. This statistic bears out what the members of the baby boom generation have said about becoming entrepreneurs rather than embracing traditional retirement (see Chapter 2). Those with at least a bachelor's degree represented 32.7% of the self-employed in 2000, but that percentage jumped to 36.6% by 2007. Self-employed Americans with a high school diploma or less dropped from 39.7% in 2000 to 36.4% in 2007. This may indicate that the work done by the self-employed is shifting from manual labor to skilled work requiring education and training. Last, there are more self-employed in urban than in rural areas (SBA, 2009).

It might be expected that levels of self-employment would decrease during recessionary times, but statistics indicate that they are not strongly affected by economic trends. Unincorporated self-employment decreased between 2007 and 2008, but incorporated self-employment stayed steady. The number of people starting businesses even increased in 2008 (SBA, 2009). It may be this willingness to take risks and innovate that helps the economy recover and thrive.

A freelance worker is somebody who is self-employed and is not committed to a particular client long term ("Freelancer," n.d.). According to a mar-

ket-research firm IDC, there were around 12 million full-time, home-based freelancers and independent contractors in America alone at the end of 2010, a
number expected to grow to 14 million by 2015. A 2011 *Bloomberg Business
Week* article notes that about $100 million worth of work was posted on Elance,

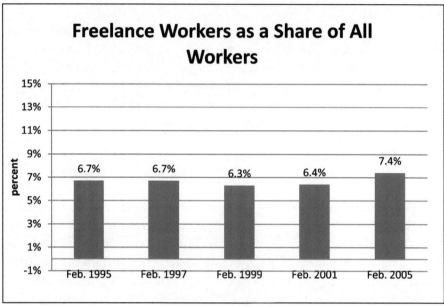

Source: U.S. Bureau of Labor Statistics, "Independent Contractors in 2005," retrieved from
http://www.bls.gov/opub/ted/2005/jul/wk4/art05.htm.

a professional freelancing company in 2010. Out of the $24.5 million that
online freelancers earned on Elance in 2010's third quarter, the largest share
went to workers in India, followed by those in the United States, Ukraine,
Pakistan, and Russia (King, 2011).
Microwork is a series of small tasks that have been broken out of a larger project and can be completed via the Internet. Think of microwork as the smallest unit of work in a virtual assembly line. The term was developed in 2008 by
Leila Chirayath Janah of Samasource, which specializes in breaking down digital work into small units and outsourcing it to the world's poor ("Freelancer,"
n.d.; "Microwork," n.d.).

Pfizer, Microsoft, Mattel, and Allstate are among companies that use
"microworkers" in place of employees, according to a *Bloomberg Business Week*
special report (King, 2011). This puts increased pressure on Western workers

through increased competition, the article explains, because businesses can easily find labor in developing markets. In 2010, the number of Web users surpassed 2 billion, of whom 1.2 billion were in the developing world, according to the ITU, the United Nations agency for information and communications technology. About 95% of people in developing countries live on less than $10 a day, according to a 2008 World Bank report.

The implications of microwork are significant for workers. Lifelong employment is no longer guaranteed. Individuals now compete in a *global* pool of talented, skilled, and educated workers. Previously education was considered the ultimate achievement in one's life, but given the new reality of globalization, workers must continually improve their education and skills so that they can remain relevant and compete in this broader talent pool.

Summary

The global market is ubiquitous and here to stay. Technology and a surge in qualified, skilled labor in developing countries have set the stage for a distributed worldwide workforce. Small businesses owners, who come from a variety of socioeconomic and demographic backgrounds, are learning to adjust to changing times through a combination of innovation and entrepreneurship—both critical to the functioning of the U.S. economy. Women are finding that starting a business of their own provides them the flexibility they need to manage other aspects of their lives. Furthermore, the rapid pace of technological integration in small business operations makes it more likely in the future that entrepreneurs will need to have special skills and training to be globally competitive. For solo workers, freelancing and microwork are growing in popularity because they can offer sustained employment. These types of work create opportunities as well as challenges for individual workers, who must now compete for jobs in a global marketplace.

References

Asherman, I., Bing, J. W., & Laroche, L. (2000). Building trust across cultural boundaries. Retrieved from http://www.itapintl.com/facultyandresources/articlelibrarymain/buildingtrust.html

Bjørn,P., & Ngwenyama, O. (2009). Virtual team collaboration: Building shared meaning, resolving breakdowns, and creating translucence. *Info Systems Journal, 19*, 227–253.

Bnet.com. (2005, April 11). Studying McDonald's abroad: Overseas branches merge regional preferences, corporate directives. Retrieved from http://findarticles.com/p/articles/mi_m3190/is_15_39/ai_n13649042/?tag=mantle_skin;content

Brannen, C., & Wilen-Daugenti, T. (1993). *Doing business with Japanese men: A woman's handbook*. Berkeley, CA: Stone Bridge.

Business Wire. (2011, June). IBM shifts global procurement headquarters to China. Retrieved from http://findarticles.com/p/articles/mi_m0EIN/is_2006_Oct_12/ai_n27040165/

Center for Women's Business Research. (2009). *The economic impact of women-owned businesses in the United States*. Retrieved from National Women Business Owners Corporation website: http://www.nwboc.org/media/CFWBR%20report%20Economic%20Impact%20WBO.pdf

Central Intelligence Agency. (2011). The world factbook: The United States. Retrieved from https://www.cia.gov/library/publications/the-world-factbook/geos/us.html

Cisco India Overview. (n.d.). Retrieved from http://www.cisco.com/web/IN/about/company_overview.html

Crimson Consulting. (2009, March). *Study shows Cisco TelePresence™ delivers rapid ROI and unique business benefits. Research Brief*. Retrieved from http://www.cisco.com/en/US/prod/collateral/ps7060/ps8329/ps8330/ps9599/TelePresence_Research_Brief_Final_03_20_09.pdf

Earle, T. C. (2009). Trust, confidence, and the 2008 global financial crisis. *Risk Analysis, 29*(6), 785–792.

The *Economist* Intelligence Unit. (2008). *The role of trust in business collaboration: An Economist Intelligence Unit briefing paper*. Retrieved from http://graphics.eiu.com/upload/cisco_trust.pdf

Edmundson, A. (2009, April). Culturally accessible E-learning: An overdue global business imperative. *T & D*. Retrieved from http://www.astd.org/LC/2009/0509_edmundson.htm

Fiol, M. C., O'Connor, E. J. (2005). Identification in face-to-face, hybrid, and pure virtual teams: Untangling the contradictions. *Organization Science, 16*(1), 19–32.

Freelancer. (n.d.). In *Wikipedia*. Retrieved June 21, 2011, from http://en.wikipedia.org/wiki/Freelancer

Hansen, F. (2006, December). Balancing the global workforce. *Workforce Management, 85*(23), 44–67.

Heitner, K. L., & Miller, L. (2011). *The great divide: Worker and employer perspectives of current and future workforce demands*. University of Phoenix. Retrieved from The University of Phoenix Research Institute website: http://www.phoenix.edu/research-institute/publications/2011/01/worker-and-employer-perspectives-of-current-and-future-workforce-demands.html

Institute for the Future, & The University of Phoenix Research Institute. (2011). *Future Work Skills 2020*. Retrieved from The University of Phoenix Research Institute website: http://phoenix.edu/content/dam/altcloud/doc/research-institute/future-skills-2020-research-report.pdf

International Monetary Fund. (2008, May). Globalization: A brief overview. Retrieved from http://www.imf.org/external/np/exr/ib/2008/053008.htm

Kandel, W. A. (2011, January 18). *The U.S. foreign-born population: Trends and selected characteristics*. Retrieved from http://www.fas.org/sgp/crs/misc/R41592.pdf

Kauffman Foundation. (2009, April 30). Entrepreneurship remains strong in 2008 with increasing business startups, according to Kauffman Foundation [Press release]. Retrieved from http://www.kauffman.org/newsroom/entrepreneurship-remains-strong-in-2008.aspx

Kauffman Foundation. (2010a). *Kauffman fast facts: Entrepreneurship and the economy*. Retrieved from http://www.kauffman.org/uploadedFiles/FactSheet/entrep_and_economy_fast_facts.pdf

Kauffman Foundation. (2010b, May 20). Despite recession, U.S. entrepreneurial activity rises in 2009 to highest rate in 14 years [Press release]. Retrieved from http://www.kauffman .org/newsroom/despite-recession-us-entrepreneurial-activity-rate-rises-in-2009.aspx

King, R. (2011, February 1) Meet the microworkers. *Bloomberg Business Week Special Report*. Retrieved from http://www.businessweek.com/technology/content/jan2011/tc20110131 _021287.htm

Klein, P. A. (2009, March–April). It's 9:05 A.M.: Do you know where your workforce is? *The Conference Board Review*, 55–59.

Kuratko, D. F. (2007, Summer). Entrepreneurial leadership in the 21st century: Guest editor's perspective. *Journal of Leadership & Organizational Studies, 13*, 1–11. Retrieved from http://www.entrepreneur.com/tradejournals/article/165018114.html

Latusek, D., & Gerbasi, A. (2010). *Trust and technology in a ubiquitous modern environment: Theoretical and methodological perspectives*. Hershey, PA: IGI Global.

McArdle, M. (2010, October). The bright side. *The Atlantic*, 46–52.

Meister, J. C., & Willyerd, K. (2010). *The 2020 workplace: How innovative companies attract, develop, and keep tomorrow's employees today*. New York: HarperBusiness.

Melting pot. (n.d.). In *Wikipedia*. Retrieved June 21, 2011, from http://en.wikipedia .org/wiki/Melting_pot

Meyer, C. (2010a, March 8). Manufacturers seek out qualified workers. Retrieved from The University of Phoenix Knowledge Network website: http://www.phoenix.edu/uopx-knowl-edge-network/articles/industry-viewpoints/manufacturers-seek-out-qualified-workers.html

Meyer, C. (2010b). Women entrepreneurs may help jumpstart the economy. Retrieved from The University of Phoenix Knowledge Network website: http://www.phoenix.edu/uopx-knowledge-network/articles/industry-viewpoints/women-entrepreneurs-help-jumpstart-economy.html

Microwork. (n.d.). In *Wikipedia*. Retrieved June 21, 2011, from http://en.wikipedia.org /wiki/Microwork

Monalisa, M., Daim, T., Mirani, F., Dash, P., Khamis, R., & Bhusari, V. (2008, July–August). Managing global design teams. *Research Technology Management, 51*(4), 48–59.

Port, D. (2010, August). But where is the money? *Entrepreneur*, 85–89.

Purdue University Center for Advanced Manufacturing. (2005). *Impact of information technology on global business*. Retrieved from http://globalhub.org/resources/2704/download/003 _Impact_of_IT_on_Global_Business_and_Leaders.pdf

Robert, L. P., Jr., Dennis, A. R., & Hung, Y.-T. C. (2009, Fall). Individual swift trust and knowledge based trust in face to face and virtual team members. *Journal of Management Information Systems, 26*(2), 241–279.

Rouse, R. A., & Cline, H. M. (2011). *Traditional and nontraditional students: Is a bachelor's degree worth the investment?* Phoenix, AZ: University of Phoenix Research Institute.

Salad bowl theory. (2009). Retrieved from http://www.docstoc.com/docs/4580772/Salad-bowl-theory

Small Business Administration. (2009). *The small business economy: A report to the president*. Washington, DC: U.S. Government Printing Office.

Shreshta, L., & Heisler, E. J. (2011, March). *The changing demographic profile of the United States. Congressional Research Service*. Retrieved from Federation of American Scientists website: http://www.fas.org/sgp/crs/misc/RL32701.pdf

Startup. (n.d.). In *Wikipedia*. Retrieved June 21, 2011, from http://en.wikipedia.org/wiki/Startup

Tan, H. H., & Lim, A. K. H. (2009). Trust in coworkers and trust in organizations. *The Journal of Psychology, 143*(1), 45–66.

Tariquea, I., & Schuler, R. (2008, August). Emerging issues and challenges in global staffing: A North American perspective. *The International Journal of Human Resource Management, 19*(8), 1397–1415.

University of Phoenix. (2011, April 14). Future Work Skills 2020. Retrieved from http://www.phoenix.edu/research-institute/publications/2011/04/future-work-skills-2020.html

Wadhwa, V., Aggarwal, R., Holly, K. Z., & Salkever, A. (2009). *The anatomy of an entrepreneur: Family background and motivation*. Retrieved from Kauffman Foundation website: http://www.kauffman.org/uploadedFiles/ResearchAndPolicy/TheStudyOfEntrepreneurship/Anatomy%20of%20Entre%20071309_FINAL.pdf

Webex. (2004). *JDS Uniphase: Enterprise edition creates value across a global enterprise*. Retrieved from http://www.webex.com/pdf/casestudy_jds.pdf

Wilen, T. (2000a). *American high technology businesswoman's strategies for working with Mexican businessmen in Mexico*. San Francisco, CA: Golden Gate University.

Wilen, T. (2000b). *International business: A basic guide for women*. Philadelphia: Xlibris.

Wilen-Daugenti, T. (2000). *Mexico for women in business*. Philadelphia: Xlibris.

Wilen-Daugenti, T. (2007). *China for businesswomen: A strategic guide to travel, negotiating, and cultural differences*. Berkeley, CA: Stone Bridge.

Wilen-Daugenti, T. (2009). *.edu: Technology and learning environments in higher education*. New York: Peter Lang.

Wilen-Daugenti, T., & Wilen, P. (1995). *Asia for women on business: Hong Kong, Taiwan, Singapore, South Korea*. Berkeley, CA: Stone Bridge.

Yakovleva, M., Reilly, R. R., & Werko, R. (2010). Why do we trust? Moving beyond individual to dyadic perceptions. *Journal of Applied Psychology, 95*(1), 79–91.

SECTION III

TECHNOLOGY TRENDS

While most people take the Internet for granted these days, the plethora of inventions and tools that have spawned from Internet technology over the decades has unleashed a whole new way of communicating, learning, interacting, and interfacing at both virtual and real-life levels. This section provides an in-depth look at technological innovations that have reinvented the way we communicate, collaborate, and engage. We discuss mobile, social, and collaborative technologies, as well as developments in presence, immersion, and gaming—both how people use them on a day-to-day basis, and the impact this use will have on higher education decision-making on delivery and content. We also look at associated factors such as impact, choice, and trust when technology is utilized for higher education purposes. The race is on to find the best way to harness interactive and learning technologies to match the needs and resources of modern consumers of higher education.

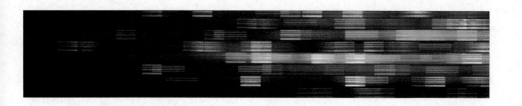

· 5 ·

MOBILE TECHNOLOGIES

Technology, which is indispensable to modern life, is constantly changing, but it is in mobile technology—both the hardware and the uses we find for it—where the biggest transformations have occurred. While the Internet made us a global community, in many ways mobile technology is allowing us to become a connected society on the move. *Mobility* used to mean the capacity for physical, geographical, or social movement or transfer. In today's usage, however, to be mobile refers to the capacity to stay connected to an electronic data source or network through wireless technology, virtually anywhere, anytime.

Mobile technology is ubiquitous and worldwide. Seventy percent of the world's population has a mobile phone—that's over 5.3 billion mobile subscribers. In America alone, 9 in 10 people own mobile phones. Children in the U.S. now are more likely to own a mobile phone than a book. Apple has sold almost 60 million iPhones worldwide, while Google's Android is growing at 886% per year. Forty-five percent of all workers report doing some or all of their work at home or by mobile devices ("Global Mobile Statistics," 2011). Almost 80% of small businesses rely on mobility as an aspect of the workplace; almost 100 million workers telecommute, and this total is rising 400% per year. The number of Americans whose employer allows them to work remotely at least one day per month increased from 7.6 million in 2004 to 12.4 million in 2006, or 63 percent (Telework Advisory Group, 2011).

The personal computer (PC) has been popular for some time, but today we are rapidly shifting from a PC-centric to a mobile-centric world. This revolution is expected to significantly influence the way we consume and disseminate information from work and at home, and for education. According to a recent projection showing that the number of smartphones, tablets and app-enabled devices will outpace PC shipments globally (Deloitte, 2011), the PC-centric world is slowly being eroded.

Given these statistics, mobility is a clear trend in society, work, and higher education. Its ease of use and availability across multiple platforms will lead mobile usage to evolve in complex ways. From the educational perspective, students increasingly expect to use their devices as part of the learning process.

Mobile Devices and Applications

Email and mobile phones have altered the nature of work and communication. Twenty years ago few people had heard of email, let alone used it, and only a small population had the cumbersome mobile phones available at the time. Less than a decade later, email had become an integral part of most workplaces, schools, and homes, and cell phones began to proliferate. Between 1990 and 2000, U.S. cell phones subscriptions rose from 5,283,055 to 109,478,031. By 2008 that number had jumped to 262,700,000 (infoplease.com). People also began to access their email and the Internet on their cell phones. In April 2009, 56% of Americans reported accessing the Internet using a mobile device, compared to only 24% in December 2007 (Horrigan, 2009).

Texting, SMS messaging, video and image swapping, and other data-driven activities are now rapidly replacing the conventional method of talking over the phone. Mobile data traffic in 2010 was over three times greater than total global Internet traffic in 2000 and is projected to increase exponentially over the next five years. It is predicted that the mobile-only Internet population will grow 56-fold, from 14 million at the end of 2010 to 788 million by the end of 2015 (Cisco, 2010).

While in previous years the Internet created unprecedented access to information and services, mobile devices today are fueling a global revolution of "off-grid, on-Net" users, who are reinventing traditional ideas of communication and human interaction. Mobile technologies are constantly in a state of flux and evolution. The important ones examined here, however, have ushered in profound changes in society, the workplace, and higher education.

Cell Phones

The shift from a dependence on telephone landlines to mobile or cell phones marked the beginning of the 24/7 connectivity era. In 2004, 65% of Americans owned cell phones; by 2008 that had risen to 78%. Cell phone ownership in the U.S. leveled out between 2009 and 2010, but the nature of cell phone use had changed significantly. The most popular activity was taking pictures, which increased from 65% of users in 2009 to 76% in 2010. Cell phone users were also more likely to send text messages, play games, exchange email, access the Internet, listen to music, and record videos on their phones than they had been the year before. In 2010, 54% of U.S. adults used their cell phones to send pictures, 23% accessed social networking through them, and 20% watched videos on them (Smith, 2010).

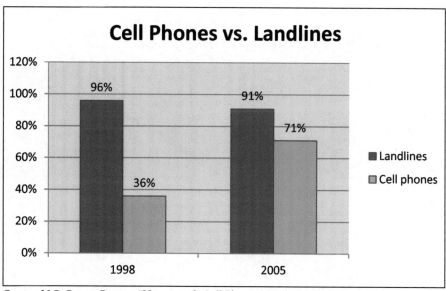

Source: U.S. Census Bureau, "Homes with Cell Phones Nearly Double in First Half of Decade," 2009, retrieved from http://www.census.gov/newsroom/releases/archives/income_wealth/cb09–174.html

Although most users do not access the Internet exclusively through their cell phones, access via that medium is increasing. More than half of all mobile users went online with their phones every day in 2010, up from 24% in 2009. Forty-three percent of those surveyed noted they go online by phone several times each day, and only 12% said they did so only once per day (Smith, 2010). This growing trend of Internet cell phone access suggests that mobility is an important factor in conducting business and getting information regardless of place and time.

Smartphones, PDAs, and Tablet Computers

A smartphone is a mobile phone that offers more advanced computing ability and connectivity than a cell phone. It allows the user to run and multitask with applications such as email, Internet, photos, video, text, and games. Smartphones combine the functions of a camera phone and a personal digital assistant (PDA). According to an early 2011 report by Olswang, a U.K.-based media and communication law firm, smartphones are experiencing accelerated rates of adoption: 22% of consumers already have one, with this percentage rising to 31% among 24–35 year olds ("Smartphone," n.d.).

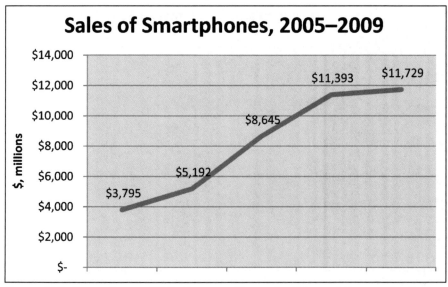

Source: U.S. Census Bureau, "U.S. Consumer Electronics Sales and Forecasts by Product Category: 2005 to 2009," 2009, retrieved from http://www.census.gov/compendia/statab/2010/tables/10s0999.pdf.

Important features of smartphones include an operating system that allows applications to be run, Web access using 3G or 4G mobile broadband networks, and wireless (Wi-Fi) support for hands-free use. Mobile broadband offers a range of high-speed connectivity through wireless cell phone carriers for accessing the

Web and data use. Popular smartphones on the market include Apple Inc.'s iPhone, Motorola's Droid, and Research In Motion's Blackberry. According to a recent Nielsen report, as of December 2010, 31% of U.S. mobile phone owners had a smartphone. eMarketer projects that smartphone ownership will reach 43% of the U.S. mobile population by 2015 ("The Future of Smart Mobile Devices," 2011).

Tablet computers are flat-screen personal computers with a touchscreen. Tablets are more compact than PCs, their main attraction being low weight, portability, and application-rich capabilities. The iPad, a tablet computer designed and developed by Apple, has generated much interest in recent years as a high-end platform for audiovisual media, including books, movies, music, and games. Microsoft, HP, Toshiba, and Dell also offer tablet computers.

The overall goal for manufacturers of tablet computers is to make them as compact as possible, with increased processing speeds and extended battery life—all highly desirable features for a multitasking population that is constantly on the move.

Laptops

Laptops are desktop replacement computers that provide the capability of these larger systems in a more mobile package ("Desktop Replacement Computer," n.d.). Just over 52% of Americans own laptop computers, a dramatic jump from 2006 when 36% of Americans owned laptops (Smith, 2010). Today laptops are a mainstay of homes and businesses, and as with tablet computers, the goal for most manufacturers is to keep up with consumer demands for reduced device size while increasing their battery life and processing speed.

The Netbook, which is a smaller version of the laptop, was introduced in 2007 to enhance their convenience aspect. Often featuring less onboard memory and storage than a laptop computer, Netbooks are sometimes coupled with the concept of cloud computing—the ability to use remote servers to free up hard-drive space on laptops, which is fast evolving as the answer to boosting speed and making processing capabilities more efficient (for more on cloud computing, please see Chapter 6). Some technology analysts now debate whether iPads and tablets have the potential to replace PCs. Most agree, however, that price and the type of activities users most want to perform on the devices will determine consumer choice (Mossberg, 2010).

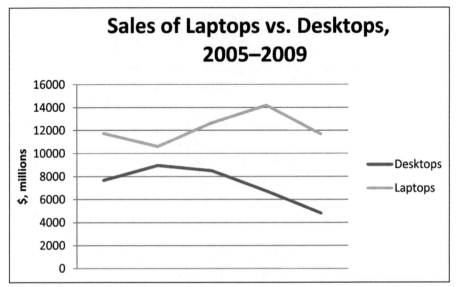

Source: U.S. Census Bureau, "U.S. Consumer Electronics Sales and Forecasts by Product Category: 2005 to 2009," 2009, retrieved from http://www.census.gov/compendia/statab/2010/tables/10s0999.pdf.

Digital Multimedia Players

Portable audio/digital devices include MP3 players, which hold audio and/or video content and provide storage for digital playback ("MP3," n.d.). The most prominent example is Apple's iPod, a small, handheld gadget that can hold close to 10,000 songs, whether burned from CDs or purchased from Apple's iTunes store. In addition to storing and playing back music, portable digital media players can also receive multimedia digital files through the Internet, whether as direct downloads or streamed webcasts to which a user subscribes. The term *podcasting* became popular sometime around 2005 when *The Guardian* newspaper coined the new word by fusing Apple's iPod with the concept of broadcasting. Podcasting allows individuals to automatically download media files to be viewed or heard while on the go or later at their convenience. ("Podcast," n.d.).

E-Readers

Wireless access, increased processing power, and better viewing quality have made handheld electronic readers (e-readers) more popular than ever. E-readers are portable, battery-powered gadgets that can store multiple books digitally and display them on an integral e-paper screen. About 5.9 million people own an e-reader in the United States, compared to 2.1 million in 2009 (Mickey, 2010). Major book retailers have released dedicated e-readers, such as Amazon's pioneering Kindle and Barnes & Noble's Nook. More than 1.8 million free books are also available (Kennedy, 2010). Electronics manufacturers offer tablets or readers with digital-bookstore access (Sony's Reader, Apple's iPad). Additionally, e-book makers have offered applications that permit users to purchase and read texts on laptop computers and smartphones.

'Apps'

Today there are thousands of applications (called "apps") that can be used on mobile devices to access and organize information away from the workplace and school. One international consulting company projects that the 7 billion apps downloaded in 2009 will grow to 50 billion by 2012, becoming a $17.5 billion industry. App stores are increasing at an incredible rate as well. The number jumped from 8 to 38 in 2009 alone. Apple has the current edge in app sales, with more than 250,000 apps (Pew Research Center, 2010) available for its iPhone, but other companies are jumping into the competition, including techno-giants Microsoft and Google. In early 2010, 59% of smartphone owners and almost 9% of feature phone users surveyed had downloaded a mobile app in the past 30 days ("The State of Mobile Apps in US," 2011). One survey cites music and cell phone ringtones as the most popular apps downloaded, followed by other forms of entertainment (Oracle, 2010).

According to a Nielsen Company survey (as cited in the Pew Project), the most popular apps for mobile devices range from (in order of importance): games, news/weather, navigation, social networking, music, entertainment/food, banking and food, sports, productivity, shopping and retail, shopping/retail, video/movies, communication and travel/lifestyle (Pew Internet and American Life Project, 2010).

Apps are being used in some markets for mobile payments. Considered an emerging technology, mobile payment taps into people's desire for convenience, and its acceptance will likely rise in the not-so-distant future. Mobile payments on smartphones are enabled by near-field communication technology (NFC), which is anticipated to be included in the next generation of smart devices from Apple, Nokia, and Google. Google Wallet, launched in 2011, uses NFC to allow consumer's smartphones to communicate with a retail

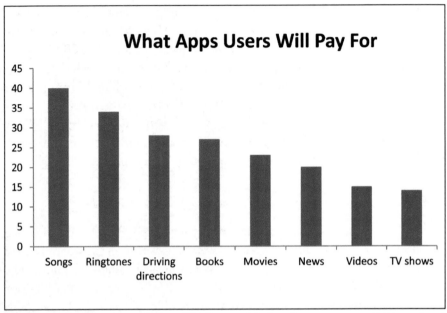

What Apps Users Will Pay For

Source: Oracle, Opportunity Calling: The Future of Online Communication, 2010, retrieved from http://www.oracle.com/us/industries/communications/oracle-communications-mobile-report-170802.pdf.

store's payment technology (Silverstein, 2011). By contrast, Starbucks has developed its own e-payment method, in which consumers buy a prepaid loyalty card with a barcode, which a smartphone reads to execute the transaction (Deloitte, 2011).

Apps are also emerging in the education sector. The University of Phoenix, the nation's largest private university, is offering its students a new way to participate in the online classroom with the PhoenixMobile App for the iPhone and iPod Touch. Launched in April 2011, the PhoenixMobile App goes beyond the campus maps or lecture downloads available from many higher education institutions, to enable University of Phoenix students to move seamlessly between the online classroom and their mobile phone. Students can now turn "on-the-go time" into learning time by participating in classroom discussions, completing assignments, and even being alerted the moment their grades are posted ("University of Phoenix Launches," 2011).

According to University of Phoenix's Chief Technology Officer, the use of apps will make education more personalized, social, and accessible to students on the go. The majority of University of Phoenix students are working learners—balancing school along with job and family responsibilities—and they do most of their school work in the online classroom between 9 P.M. and 2 A.M. (University of Phoenix, 2011).

Mobility in Society

By 2010 the picture of mobile technology had changed dramatically from just a few years before. A study done in May 2010 looked at how Americans were using mobile technology (defined as laptops with Wi-Fi or mobile broadband) for Internet access, instant messaging, and using apps on mobile phones. The survey found that 59% of adults in the U.S. now go online wirelessly, up from 51% in 2009 (Smith, 2010). Mobile devices provide people with more flexibility on where they can connect to the Internet. Among laptop owners, 86% said they go online from home, 37% from work, 37% from other locations, and 29% from all three (Smith, 2010).

Device Ownership

Eighty-five percent of Americans own cell phones (Pew Research Center, 2011). Three quarters own either a desktop or laptop PC. MP3 players and game console adoption is on the rise, while e-readers and tablet computers—now owned by less than 10% of Americans—are rapidly catching up.

Other demographics show that a younger male population dominates the gadget-buying industry, with affluence and education both important factors in the sales of e-readers and iPads. In a 2010 survey of 5,000 owners of mobile devices such as game players, e-book readers (like the Kindle and the Nook), tablet computers, media players, smartphones, and iPods, 65% of iPad users were male, and 63% were under 35. Kindle owners tended to be more affluent, with 44% making more than $80,000 per year. They were also better educated; 27% had a master's or doctoral degree. In comparison, only 39% of iPod users and 37% of iPhone users made over $80,000 a year (Nielsen, 2010).

Older Americans are catching up to younger generations in use and ownership of some electronic devices. A recent Pew Center survey of 3,001 adults showed that 85% of them owned cell phones, and 90% lived in a household with at least one mobile phone. Despite the proliferation of laptops, the desktop computer is more popular with adults between 35 and 65 years, as opposed to younger adults, 70% of whom owned laptops (Zickuhr, 2011).

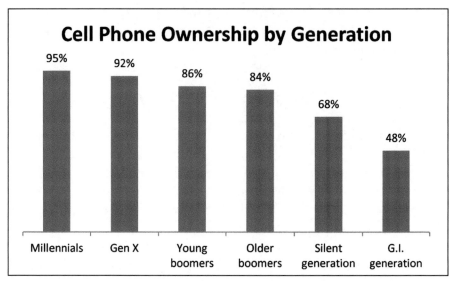

Source: The Pew Internet and American Life Project, "A Closer Look at Generations and Cell Phone Ownership," 2011, retrieved from http://pewinternet.org/Infographics/2011/Generations-and-cell-phones.aspx

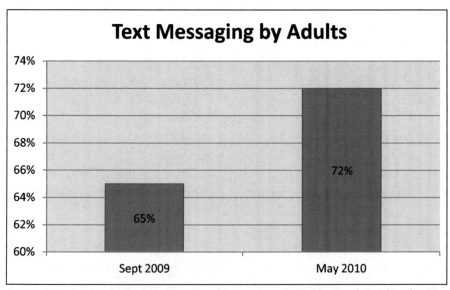

Source: U.S. Centers for Disease Control and Prevention, *CDC Data Brief, Social Media: Text Messaging,* retrieved from http://www.cdc.gov/healthmarketing/ehm/databriefs/mobiletext.pdf

In terms of other mobile devices, older Americans tend to lag behind considerably from the younger Millennial generation (18–34 years). Most adults surveyed used two of the non-voice functions on their cell phone—text messaging and taking pictures—while Millennials used a variety of features, such as Web surfing, emailing, listening to music, and recording videos.

Mobile Device Uses

According to a Pew Center for Research survey (Rosentiel & Mitchell, 2011), Americans use mobile devices to access practical information in real time. For example, 42% of mobile device owners report using them to receive weather updates, and 37% research restaurants or other local businesses on their phones or tablets.

Use of apps is another growing trend among owners of mobile devices. Thirteen percent of all mobile device users report accessing local information via an app. Many apps are available for free, but 10% of mobile-device app users say they pay for the ones that connect them to local news and information. Accessing online health information is also rising in popularity with mobile device owners. According to a 2010 Pew Center survey, 17% used their cell phones to look up health and medical information (Fox, 2010). Young cell phone users between the ages of 18 and 29 were more likely to use these apps, which range from calorie counters, sources of nutrition information, body mass index calculators, personal health record organizers, and daily prompts for medication and exercise routines.

As to the age demographics of mobile device users, younger adults use technology more often than their older counterparts, although the 30–49 age group is starting to catch up. For example, 95% of people in the 18–29 age group send text messages on their phones, and 65% access the Internet with mobile devices. Nineteen percent access the Internet only through their cell phones. Among adults 30–49, 83% used them to take pictures in 2010, up from 71% the year before; 39% record videos, up from 21%; and 35% use instant messaging, up from 21%. Thirteen percent access the Internet exclusively by cell phone (Smith, 2010). Younger users use their phone as an entertainment device and often as a mini-computer (Oracle, 2010). The accelerated pace of gadget ownership suggests that mobile devices are creating an untethered lifestyle, which indicates no signs of abating any time soon.

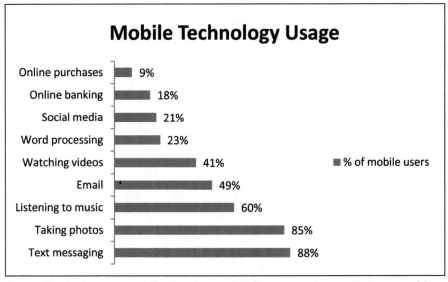

Source: *Oracle, Opportunity Calling: The Future of Online Communication*, 2010, retrieved from http://www.oracle.com/us/industries/communications/oracle-communications-mobile-report-170802.pdf.

Recent U.S. Census Bureau data show that older adults are actively using the Internet. They indicate that 42% of adults over 65 and older access the Internet for sending and reading e-mail and use a search engine to find information and get news online. However, there is no clear picture yet of the degree to which this types of information is accessed using Internet-enabled mobile devices.

Future Mobile Device Trends

Among the latest mobile technologies, e-readers seem to be the wave of the future. The market for e-books took off in 2010. The Yankee Group predicts that this market will go from about $1.3 billion in 2010 revenue to as much as $2.5 billion in 2013 (Wireless Week, 2010a). The prices of e-readers will decrease by an average 15% per year for the next five years, and the Yankee Group predicts that the price decrease will produce a 55% increase in use of the machines each of those years. Strategy Analytics is predicting that Apple's iPad and tablets produced by other manufacturers will change the market category, since they can be used as e-readers and for many other purposes. Strategy Analytics estimates that the new tablet category will be worth over $11 billion by 2014 (Wireless Week, 2010a).

According to the American Library Association, 66% of libraries are now lending e-books, compared with 38% in 2005 (Grundner, 2010). E-reader owners also report a higher appetite for books. A survey by Marketing & Research Resources, Inc. revealed that 40% of e-reader owners were reading more with e-readers than they did with printed books. The average e-reader owner reads 2.6 books per month, according to the study, versus 1.9 books for print-only readers (Grundner, 2010).

Device ownership is on a steep upward trajectory in the U.S. that shows little signs of slowing down, especially as mobile devices increasingly integrate multiple features on a single unit. Part of the difficulty in predicting market growth for new mobile technologies is that devices are constantly being developed that can replace whole categories, in the way tablet computers may wipe out e-book readers. Researchers are developing items that would have looked like science fiction only a few years ago, such as electronic devices made from a stretchable material that can be worn on the back of the hand or on the wrist à la Dick Tracy or Babylon 5 (Levine, 2008) and flexible electronic displays for computer screens, e-papers, calculators, and charts (Horizon Report, 2010). Similarly, combining social networking capabilities (discussed in more detail in Chapter 6) with mobile devices is expected to create the next-generation platform for launching unprecedented advances in communication and commerce (MacManus, 2009).

In global terms, networking giant Cisco predicts not only an exponential uptick in global data traffic and mobile devices worldwide, but also a period when mobile networks will eventually break the electricity barrier in parts of Asia and Africa (i.e., more people will have access to mobile networks than to electricity at home; Cisco Systems, 2011). In the Asia-Pacific region alone, for instance, there were nearly 678 million mobile communications subscribers in 2005 (Frost & Sullivan, 2006). South Korea had the world's highest percentage of users of third-generation mobile phones in 2007 and had a market penetration of 93% (Jung & Leckenby, 2007). Mobile phone use has also skyrocketed in China, the world's largest device market. The Chinese are more likely to access the Internet through cell phones (38%) than Americans (27%). With China's huge population, this translates to 755 million cell phone subscribers in 2010. The biggest group of users is adults age 24–44, possibly because many younger people cannot afford cell phone charges. Aside from making calls, Chinese users most commonly use their phones for texting (87%; Philips, 2010).

Sustained growth of mobile technologies in society thus seems inevitable and will significantly impact the way we communicate and manage our daily lives. Mobile technologies have the potential to provide the flexibility that is so greatly needed in the modern workplace. Although they can be essential in helping people balance their work and family lives, it's also possible for such technologies to cause them to overlap, making a balance more difficult to achieve by denying users sufficient time to focus solely on family and leisure. Learning to use this technology wisely to create a vibrant economy and a happy and healthy workforce is the current challenge of mobile technology.

Mobility in Work

Mobile devices are allowing employees to work outside the office more often than ever before. As discussed earlier, telecommuting has become an attractive option for many companies and workers as a way to incorporate flexibility and improve productivity in the workplace. In 2008, 53% of American adults were employed either as full- or part-time workers, of whom 62% were "networked workers," defined as those who used the Internet, email, cell phones, computers, PDAs, and other devices to work in a variety of locations (Madden & Jones, 2008). These workers are more likely to own mobile devices than the typical U.S. adult. For example 93% of networked workers owned cell phones, compared to 78% of adults overall. Sixty-one percent used laptops, compared to 39% of all adults; and 27% had PDAs, versus 13% of all adults. Overall, cell phone ownership among employed Americans is rising: 89% in 2008, versus 82% in 2006 (Madden & Jones, 2008).

Working from Home

The work-from-home trend has been growing steadily as a direct outcome of mobile devices and broadband Internet access. By 2008, according to a Pew Internet & American Life Project study, 45% of employed Americans were doing some work at home, while 18% worked at home every or almost every day. A higher percentage (20%) of networked workers worked at home every day or almost every day. Of the employed adults who worked at home, 37% worked at home at least two times a month, while 56% of networked workers did (Madden & Jones, 2008).

In 2008, 60% of workers said they used the Internet at work every day, while 27% said they used it constantly. Some groups of workers used it more than others. For example, 73% of workers in government and educational institutions said they used the Internet several times a day, while almost 75% of professionals, managers, and executives said they used the Internet either constantly or several times a day. That compares to rates of around 50% for clerical and sales workers. Service workers use the Internet least of all in their work (Madden & Jones, 2008).

Ninety-six percent of working adults say they used information and communication technology in their work in some way in 2008. Broken down, 86% used email or the Internet, 89% used cell phones, and 81% had personal or work email accounts; 73% used all three (Madden & Jones, 2008).

Changes in Workforce Management

Mobile applications are changing workforce management. Aside from checking inventories and performing analytics against the corporate database, Web-enabled cell phones, smart phones, and tablet computers are being used for recruitment and other human resource functions such as scheduling and employee and manager self-service. According to J. Gerry Purdy, principal analyst at MobileTrax LLC, mobile devices are becoming strategic corporate assets (Purdy, 2011).

In a study of 230 organizations to assess the impact of mobile devices on human capital management (HCM), the Aberdeen Group showed that 53% used mobile devices for workforce management, 39% for informal learning and development, and 38% for talent recruitment. There were several factors that drove companies to use mobile devices for HCM, including needs and expectations of multiple generations (44%), economic conditions creating a drive for efficiency (40%), and a dispersed, distributed workforce (40%). Moreover, 65% of employees using mobility tools rated themselves as "highly engaged," versus 57% at organizations without mobile HCM (Lombardi, 2010)

Trends in Mobile Workers

The International Data Corporation (IDC) reported that by 2010, the U.S. had the highest percentage of mobile workers in the world at 72.2%. IDC predicted that 75.5% of U.S. workers would be mobile by 2013, a total of 119.7 million workers. Though the U.S. has the largest percentage of mobile workers, it by no means has the most. The market in the Asia/Pacific (excluding Japan) has a lower percentage of mobile workers but a much larger number of workers. IDC predicts that mobile workers in this area will increase from 546.4 million in 2008 to 734.5 million workers in 2013, which will be only 37.4% of its workforce. By 2013, this area of Asia should have 62% of the world's mobile workers (Wireless Week, 2010b).

The number of mobile workers in Western Europe and Japan is also increasing. IDC estimates that 50.3% of the Western European workforce and 74.5% of the Japanese workforce will be mobile by 2013, comprising 129.5 million and 40.3 million workers, respectively. Overall, the number of mobile workers topped 1 billion in 2010 and is expected to reach 1.2 billion by 2013 (Wireless Week, 2010b).

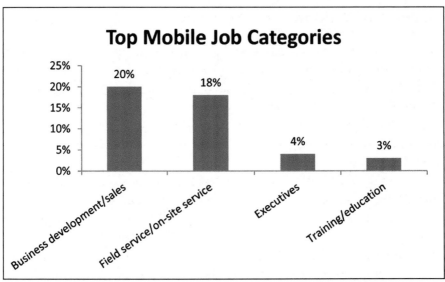

Source: "Mobile & Wireless Slideshow: Managing Mobile Workers—10 Fast Facts," by D. McCafferty, 2011, *CIO Insight*, retrieved from http://www.cioinsight.com/c/a/Mobile-and-Wireless/Managing-Mobile-Workers-10-Fast-Facts-180272/

Impact of Remote Work on Society

The proliferation of mobile devices means that more work can be done remotely. This mobility allows flexibility in the workplace but can also have adverse effects via the intrusion of work-related pressures into personal lives. Most people have viewed the rising trend of work-related Internet use, either at home or in other locations, as positive, but a significant number of commentators have noted that using these devices can also create stress (Madden &

Jones, 2008). The vast majority of workers surveyed in 2008 by the Pew Internet & American Life Project thought these devices improved their ability to do their jobs and share information with others and that the devices gave them more flexibility. Almost half of the workers, however, complained that these devices had increased demands to work more, created stress, and made it harder to disconnect their work from their home lives. People who owned PDAs experienced higher levels of stress; the Pew study noted that professionals, managers, and executives were more likely to own all gadgets, including PDAs.

The potential for increased stress is reflected in the 2008 Pew survey, which reported that 37% of employees checked their email constantly, up from 22% in 2000. In 2002, only 16% of workers checked their email often on weekends. By 2008 that picture had changed; half of the workers checked email on weekends, and 22% checked it often; 46% said they checked in on sick days, and 25% did so often; and 34% checked it on vacation, while 11% checked it often. Among PDA owners the percentages were higher for every category. Twenty-two percent of workers and 48% of working PDA users said they were required to check work-related email outside of work hours (Madden & Jones, 2008).

Clearly, mobile devices have changed how and when workers stay connected to their jobs, impacting the work/life balance that many strive for. Some researchers suggest that while communication technologies have created flexibility and convenience, they have also had many negative effects. Electronic communications allow employees to be reachable at all times, and it reduces their engagement and productivity by creating interruptions and distractions in their lives (Mindrum, 2011). Too many distractions, and the hazards of multitasking, can slash employee productivity while increasing their stress.

On the other hand, there are studies that point to the potential for less worker stress when they use mobile technologies while telecommuting. Remote workers were protected from workplace interruptions, such as office politics, meetings, and information overload, leading to fewer distractions. Another study of more than 24,000 IBM employees in 75 countries showed that workers who telecommuted had a higher threshold of tolerance for being interrupted by work in their personal lives than those who worked fewer hours in the workplace itself. In other words, telecommuting meant more productivity and personal satisfaction with work/life balance issues than working fewer hours in the office itself (Sabatini, 2010).

Mobility in Higher Education

Student life has changed substantially over the past decade. In the past, students tended to interact with those who were in the same geographical place as them. They would attend classes and discussion groups, go to the library, join campus organizations, meet with faculty, and get together with friends. Technology, however, has revolutionized much of this traditional campus environment. Today's students now use and access intelligent whiteboards, chat tools, videoconferencing systems, digitized movies, electronic libraries, and mobile devices (Lloyd, Dean, & Cooper, 2007). These technologies often mean that students spend less time meeting face-to-face with other students and the faculty. The other students and the faculty may not even be in the same physical location. Students are on the move, using laptops and cell phones to perform research, organize information, and prepare presentations from wherever they happen to be. As students rapidly adopt new technologies and the latest mobile gadget on the market, universities and educational institutions are finding that they have to stay technologically relevant in an environment increasingly driven by student needs and expectations.

Student use of mobile devices on college campuses has exploded in recent years ("How Can Smartphones Help," 2011).The site also notes, however, that currently only about 8% of higher education institutions have a central mobile presence. Harvard and Ohio State University are offering students campus-wide maps on their mobile sites. Handheld computers like smartphones and tablets have risen from being niche items to primary modes of Internet access in higher education institutions (Keller, 2011). Many innovative mobile concepts are in the making: Stanford University hopes to replace students' ID cards with electronic versions stored on their phones, and the University of Washington is holding student competitions for new iPhone applications (Keller, 2011).

There is also a growing trend for open-source content in higher education, such as MIT's open-source software (ocw.mit.edu) and Stanford's Engineering Everywhere (see.stanford.edu). Jasig, a consortium of educational institutions and commercial affiliates, is launching a new open-source project called uMobile and is calling on colleges and universities to contribute to the effort. Jasig aims to deliver projects that use standards-based integration, authentication, security, single login, and customization. The new uMobile project will be built on the uPortal framework, which is a leading open-source enterprise portal framework built by and for the higher education community. Initial applications proposed are campus maps and directories, RSS feeds, calendars, course schedules, campus news, and other tools commonly used on mobile devices (Nagel, 2011).

Student Use of Mobile Devices

U.S. students and their families spent about $12.8 billion on electronic devices to use in connection with their education in 2007, an increase of around 22% from 2006. In this they are barely catching up with their European counterparts, who have been ahead of the U.S. in the use of mobile devices for most of the last decade. For example, by 2003, 75% of the general population in the United Kingdom, and 90% of young adults there, had mobile phones (Crabtree, Nathan, & Roberts, 2003).

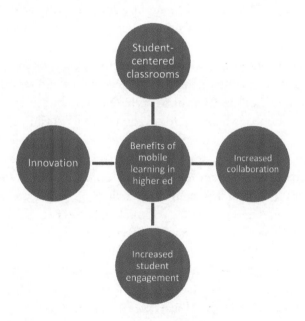

By the end of 2003, it was estimated that 82% of the population of the Netherlands used mobile devices, compared to 54% of the population in the United States at that time. At the same time, 98% of children in the Netherlands were connected using mobile devices by age 14, and 100% of young adults age 16–22 had at least one mobile device (Wentzel, von Lammermen, Molendijk, de Bruin, & Wagtendonk, 2005).

EDUCAUSE, an education-technology research organization, found that student ownership and use of technology has undergone significant change over the last four years. Where computer ownership has remained steady, the shift to laptops has been dramatic. Desktop ownership declined by 25% over the last four years with 89% of students owning either a Netbook or laptop computer (Smith & Caruso, 2010).

Many colleges and universities began facilitating student mobility several years ago by developing wireless campuses, where students could use their laptops and other mobile devices in various places across school grounds (Levine, 2002). By 2002, more than 90% of public and 80% of private universities in the United States had some mobile wireless technology (Swett, 2002). It has since become an integral part of colleges and universities. In 2001, 57% of U.S. campus libraries already had wireless networks. By 2003 that had grown to 88% (Boggs, Smolek, & Arabasz, 2002), and by 2005 students at several colleges and universities began requesting wireless access everywhere on campus (Wilen, 2005).

The rapid growth of free wireless connectivity on campus has opened up many options for how course content can be delivered and accessed, based on individual preference and learning style. Students at the University of Connecticut School of Medicine, for example, use PDAs during clinical rotations to find quick answers to treatment issues, such as possible drug interactions or best practices in patient care. They also used the devices to track patient encounters and keep a student log (Morgen & Smith, 2008). Other examples of higher education use of this technology include providing students with preloaded smartphones and pocket PCs that allow access to college communications, academic tools, and coursework (Trella & Swiatek-Kelly, 2008); data collection for later analysis; instructional audio and video file downloads to handheld devices (Higher Education Funding Council for England, 2005); and using PDAs to view study materials and answer quizzes downloaded from laptops used in the classroom (Houser & Thornton, 2004).

If used wisely, wireless technology can boost efficiency and effectiveness in learning environments (Maginnis, White, & Mckenna, 2000). This technology is also cost effective and will become more efficient as the costs decrease and the quality of wireless services continues to improve (Boggs, Smolek, & Arabasz, 2002; Galbus, 2004).

Students Engaged in M-Learning

The use of mobile devices in education has been given a name: M-learning. This is defined as the intersection of mobile computing with e-learning (Quinn, 2000), and as "any activity that allows individuals to be more productive when consuming, interacting, or creating information mediated through a compact, digital portable device that the individual carries on a regular basis. Such a device also has reliable connectivity and fits in a pocket or purse" (Wexler et al., 2007, p. 6). Devices included within this definition are mobile phones, PDAs, MP3 players, game consoles, and tablet computers.

Laouris and Eteokleous (n.d.) argue that M-Learning or learning on the move has been around for years and has now reemerged as fashionable with the proliferation of mobile devices. The researchers note that there is no single definition of M-Learning and perhaps too many. They conducted a Google search in 2005 and retrieved 1,240 definitions of M-Learning; when they searched again in June 2006, they found 22,700.

With M-learning, the classroom is no longer limited to what it used to be: a room with four walls where students and instructors meet. M-learning permits each individual student to set up a learning program and environment that suits him or her. This form of learning is not restricted to the physical classroom or a set time period. Mobile devices also help students learn in ways that education researchers say will increase their learning ability (Bestwick & Campbell, 2010).

Today mobile devices are being used to fit coursework into work or personal schedules and are helping students to better manage their time. For instance, some studies show that these devices give commuter students the ability to use spare time wisely by finishing homework and preparing for lessons (Lehner, Nosekabel, & Lehmann, 2003).

Podcasts, short online videos, and audio recordings can be downloaded onto iPods or smartphones and listened to or viewed at any time, and many colleges and universities now provide online recordings of lectures either on their own sites or through portals like YouTubeU and iTunesU. Students can review lectures or virtually attend those they missed at any time, allowing them to make good use of commute time and occasional breaks in busy schedules. iPods and iTunes also offer educational and audio recordings via the Web.

Cell phones in particular are being used for an increasing number of educational purposes, from electronic exchange of pictures and video recordings to Web searches and email and texting. Downloaded smartphone apps help students organize assignments, map out university campuses, and locate free Wi-

Fi hotspots, and many are free depending on the smartphone model being used. Among those students may find particularly useful, Evernote is an app that takes screenshots of websites to form a library of clippings. It can also be used to record random notes, voice messages, and pictures that students can store for future use in assignments. MyPocketProf syncs notes to an iPhone to be viewed offline at a later time. Stanza allows users to browse among 100,000 books and read them on their mobile phone. Many universities provide their own applications to students for free. All of these apps can be very useful for students in communicating with each other and their instructors, recording and sharing information, and looking up information online.

Potential Inhibitors to M-Learning

Students are riding the wave of mobility in unprecedented numbers, but it is important to remember that many factors can limit accessibility to mobile devices at both the individual and institutional levels. High charges for data usage and expensive monthly plans can often be economically prohibitive to some students. Similarly, a steep jump in mobile device usage on campus may mean that colleges and university enterprises must actively prepare their staff and deploy resources to ensure adequate mobile hardware and software support, e-security measures, and technical support when required (Katz, 2009). University budgets also can find the costs of maintaining technological innovations costly and of lower priority when compared to other urgent needs.

On another front, some instructors tend to view mobile devices like cell phones as more of a distraction than a teaching tool, and they have been hesitant to fully incorporate them into their teaching program. Instructors may also be much less comfortable using this technology than younger students who began using these devices earlier in life, and they may not have fully examined how to use this technology optimally in the learning environment. Furthermore, educators are concerned about security and privacy in mobile learning scenarios. Despite these obstacles, mobile devices continue to receive positive evaluations by researchers and students, and their use in education continues to grow.

Summary

Independent of time and location for their use, mobile devices have become a revolutionary communication tool for a broad sector of American society. Our social interactions, workplace dynamics, and educational environment are changing in fundamental ways due to an increasingly mobile and connected world. New and more sophisticated devices are constantly being developed as users of all ages and demographics are finding new and different uses to bring flexibility and convenience into their lives at home, in the workplace, and in education. As knowledge and information is accessed using different modalities and platforms, M-learning is also gaining interest among learners. Higher education institutions should recognize the importance of mobile technologies in peoples' lives and provide programs that cater to students' embrace of new learning tools and devices.

References

Bestwick, A., & Campbell, J. R. (2010, September). Mobile learning for all. *The Exceptional Parent Magazine*. Retrieved from http://findarticles.com/p/articles/mi_g02827/is_9_40/ai_n 56427545/

Boggs, R., Smolek, J., & Arabasz, P. (2002, June 11). *Choosing the right wireless network: A technology challenge for higher education* (EDUCAUSE Research Bulletin, Vol. 2002, Issue 12). Retrieved from EDUCAUSE Center for Applied Research website: http://net.educause.edu/ir/library/pdf/ERB0212.pdf

Cisco Systems. (2011). *Cisco visual networking index: Global mobile data traffic forecast update, 2010–2015*. Retrieved from http://www.cisco.com/en/US/solutions/collateral/ns341/ns525/ns537/ns705/ns827/white_paper_c11-520862.pdf

Crabtree, J., Nathan, M., & Roberts, S. (2003). *MobileUK: Mobile phones and everyday life*. Retrieved from http://www.theworkfoundation.com/research/publications/publication detail.aspx?oItemId=103

Deloitte. (2011). *Tech trends 2011. The natural convergence of business and IT*. Retrieved from http://www.deloitte.com/assets/Dcom-UnitedStates/Local%20Assets/Documents/us_ consulting_techtrends_021511.pdf

Desktop replacement computer. (n.d.). In *Wikipedia*. Retrieved June 21, 2011, from http://en.wikipedia.org/wiki/Desktop_replacement_computer

Fox, S. (2010, October 19). Mobile health 2010. Retrieved from http://www.pewinternet .org/Reports/2010/Mobile-Health-2010/Report.aspx?view=all.

Frost & Sullivan. (2006, June 5). Mobile subscriber base approaches one billion in Asia-Pac [Press release]. Retrieved from http://www.frost.com/prod/servlet/press-release.pag?docid= 71139176&ctxixpLink=FcmCtx1&ctxixpLabel=FcmCtx2

The future of smart mobile devices. (2011, February 10). Retrieved from http://www. emarketer.com/Article.aspx?R=1008228

Gilroy, A. (2009, August). Extra, extra, e-read all about it: More e-readers debut. *Consumer Electronics, 28*.

Global mobile statistics 2011. (2011). Retrieved from http://mobithinking.com/mobile-marketing-tools/latest-mobile-stats

Grundner, A. (2010, September 30). Kindle owners tend to be wealthy and educated, iPad own-

ers tend to be under 35 and male. Retrieved from http://www.ehomeupgrade
.com/2010/09/30/kindle-owners-tend-to-be-wealthy-and-educated-ipad-owners-tend-to-
be-under-35-and-male

Higher Education Funding Council for England. (2005). *Empowering learners: Mobile learning and
teaching with PDAs.* Retrieved from http://www.elearning.ac.uk/innoprac/practitioner
/resources/dewsbury

Horizon Report. (2011). Four to five years: Flexible displays. *2010 Horizon Report: The K12 edition* Retrieved from http://wp.nmc.org/horizon-k12-2010/chapters/flexible-displays/

Horrigan, J. (2009, July 22). Wireless internet use. Retrieved from http://www.pewinternet
.org/Reports/2009/12-Wireless-Internet-Use.aspx

Houser, C., & Thornton, P. (2004). Japanese college students' typing speed on mobile devices.
In *Proceedings. 2nd IEEE International Workshop on Wireless and Mobile Technologies in
Education* (pp. 129–133). Retrieved from http://ieeexplore.ieee.org/xpl/freeabs_all.jsp?tp=
&arnumber=1281353&isnumber=28620

How can smartphones help college students find themselves? Answer: Campus maps. (2011).
Retrieved from http://higheredlive.com/how-can-smartphones-help-college-students-find-
themselves-answer-campus-maps/

Jung, J., & Leckenby, J. (2007). *Attitudes toward mobile advertising acceptance and behavior inten-
tion: Comparison study of Korea and U.S.* Research proposal presented at the American
Academy of Advertising, Asia-Pacific Annual Conference, Korea University, Seoul.

Katz, R. N., (2009). *Higher education: A moveable feast* (Students and Information Technology
2009, ECAR Research Study 6). Retrieved from EDUCAUSE Center for Applied Research
website: http://net.educause.edu/ir/library/pdf/ers0906/rs/ers09062.pdf

Keller, J. (2011, January 23). As the web goes mobile, colleges fail to keep up. Retrieved from
http://chronicle.com/article/Colleges-Search-for-Their/126016/

Kennedy, S. G. (2010, October). Internet waves. Ebooks by the numbers. *Information Today,*
15–17.

Laouris, Y., & Eteokleous, N. (2005). We need an educationally relevant definition of mobile
learning. Retrieved from mLearn website: http://www.mlearn.org.za/CD/papers/Laouris
%20&%20Eteokleous.pdf

Lehner, F., Nosekabel, H., & Lehmann, H. (2003, June). Wireless e-learning and communication environment: WELCOME at the University of Regensburg. *e-Service Journal, 2*(3),
23–41.

Levine, B. (2008, February 25). Dick Tracy would love Nokia's Morph cell phone. Retrieved from
http://www.newsfactor.com/story.xhtml?story_id=58518&full_skip=1

Levine, L. M. (2002, October). Campus-wide mobile wireless: Mobility and convergence.
Syllabus. http://campustechnology.com/articles/2002/09/campuswide-wireless-mobility-and-
convergence-an-interview-with-lawrence-m-levine.aspx

Lloyd, J., Dean, L. A., & Cooper, D. L. (2007). Students' technology use and its effects on peer relationships, academic involvement, and healthy lifestyles. *NASPA Journal, 44*(3), 481–495.

Lombardi, M., (2010, June). *Mobile HCM: Workforce and talent management on the move.*
Retrieved from Aberdeen Group website: http://www.adp.com/solutions/employer-
services/time-and-attendance/workforcemanagement/~/media/DCD6C3
4484BF440498D74DB4AF19A5AF.ashx

MacManus, R. (2009, October 20). *Mary Meeker's Internet trends presentation 2009*. Retrieved from ReadWriteWeb website: http://www.readwriteweb.com/archives/mary_meekers_internet_trends_presentation_2009.php

Madden, M., & Jones, S. (2008, September 24). *Networked workers*. Retrieved from Pew Internet & American Life Project website: http://www.pewinternet.org/~/media//Files/Reports/2008/PIP_Networked_Workers_FINAL.pdf

Maginnis, F., White, R., & McKenna, C. (2000, November–December). Customers on the move: M-Commerce demands a business object broker approach to EAI. *eAI Journal*, 58–62.

Mindrum, C. (2011, March). The twitching organization. *Chief Learning Officer*. Retrieved from http://clomedia.com/articles/view/4128

Mickey, B. (2010, November 29). Study: Number of e-reader owners triples in under two years. *Folio*. Retrieved from http://www.foliomag.com/2010/study-number-e-reader-owners-almost-triples-under-two-years

Miller, R. (2010, May 26). Apps: exploring the next content frontier. Retrieved from http://www.econtentmag.com/Articles/ArticleReader.aspx?ArticleID=67496&PageNum=2

Mindrum, C. (2011, March). The twitching organization. *Chief Learning Officer*. Retrieved from http://clomedia.com/articles/view/4128

Morgen, E., & Smith, B. (2008, January 25). Student use of PDAs at the UConn Health Center School of Medicine. Presented at the Teaching and Learning With Mobile Technologies Event, NorthEast Regional Computing Program, Southbridge, MA.

Mossberg, W. (2010, March 31). Apple iPad review: laptop killer? Pretty close. Retrieved from All Things D website: http://ptech.allthingsd.com/20100331/apple-ipad-review

MP3. (n.d.). In *Wikipedia*. Retrieved June 21, 2011, from http://en.wikipedia.org/wiki/MP3

Nagel, D. (2011, April 25). Open source group seeks support from higher ed for mobile initiative. Retrieved from http://campustechnology.com/articles/2011/04/25/open-source-group-seeks-support-from-higher-ed-for-mobile-initiative.aspx

Nelson, M., & EDUCAUSE. (2008, March/April). E-books in higher education: Nearing the end of the era of hype? *EDUCAUSE Review*, *43*(2). Retrieved from http://www.educause.edu/EDUCAUSE+Review/EDUCAUSEReviewMagazineVolume43/EBooksinHigherEducationNearing/162677

NielsenWire. (2010, September 13). Insights on the emerging mobile app economy. Retrieved from http://blog.nielsen.com/nielsenwire/media_entertainment/insights-on-the-emerging-mobile-app-economy

Oracle. (2010). Opportunity calling: The future of mobile communication. Retrieved from http://www.oracle.com/us/industries/communications/oracle-communications-mobile-report-170802.pdf

Parry, D. (2008, March 24). Teaching with Twitter [Video file]. Retrieved from http://chronicle.com/blogs/wiredcampus/most-popular-wired-campus-tv-installments/4442

Pew Internet & American Life Project. (2010, September 14). Rise of the "apps culture." Retrieved from http://pewresearch.org/pubs/1727/cell-phone-apps—popular-download-demographics

Philips, S. (2010, August 4). Mobile Internet more popular in China than in the U.S. Retrieved from http://blog.nielsen.com/nielsenwire/global/mobile-internet-more-popular-in-china-than-in-u-s/

Podcast. (n.d.). In *Wikipedia*. Retrieved June 21, 2011, from http://en.wikipedia.org/wiki/Podcast

Purdy, J. (2011, March). Mobile workforce management? *HRMagazine, 56*(3), 67–70.

Quinn, C., & Hobbs, S. (2000). Learning objects and instruction components. *Educational Technology & Society* 3(2), retrieved from http://citeseerx.ist.psu.edu/viewdoc/download? doi=10.1.1.36.329&rep=rep1&type=pdf

Rosentiel, R., & Mitchell, A. (2011). Survey: Mobile news and paying online. Retrieved from http://stateofthemedia.org/2011/mobile-survey/

Sabatini, P. (2010, July 26). Workzone: Flexibility—A no-cost boost for morale, productivity. *The Pittsburgh Post-Gazette*. Retrieved from http://www.post-gazette.com/pg/10207/1074851 –407.stm#ixzz10jNGOojU

Silverstein, E. (2011, June 7). Google Wallet makes it easier to pay for goods through near field communications. *TMCnet.com*. Retrieved from http://www.tmcnet.com/topics/articles/ 183328-google-wallet-makes-it-easier-pay-goods-through.htm

Smartphone. (n.d.). In *Wikipedia*. Retrieved June 21, 2011, from http://en.wikipedia.org/wiki/ Smartphone

Smith, A. (2010, July 7). *Mobile access 2010*. Retrieved from Pew Internet & American Life Project website: http://www.pewinternet.org/~/media/Files/Reports/2010/PIP_Mobile_ Access_2010.pdf

Smith, S. D., & Caruso, J. B. (2010, October). *The ECAR study of undergraduate students and information technology, 2010*. Retrieved from EDUCAUSE Center for Applied Research website: http://net.educause.edu/ir/library/pdf/EKF/EKF1006.pdf

The state of mobile apps in US. (2011, April 8). Retrieved from Online Marketing Trends website: http://www.onlinemarketing-trends.com/2011/04/state-of-mobile-apps-in-us.html

Swett, C. (2002, October). College students' use of mobile wireless-Internet connections becomes more common, *Knight Rider Tribune Business News*, Washington, DC.

Symbol Technologies (2005). *Symbol Technologies helps unplug the wires at Monash Medical Centre*. Retrieved from Motorola website: http://www.motorola.com/web/Business/Solutions/ Industry%20Solutions/Healthcare/Mobile%20Physician%20Rounding/_Document/ static%20files/case-study-monash-medical-centre.pdf

Telework Advisory Group of WorldatWork. (2011). [Home page]. Retrieved from http://www. workingfromanywhere.org

Trella, J., & Swiatek-Kelley, J. (2008, January 25). Extending campus resources to the mobile device. Presented at the Teaching and Learning With Mobile Technologies Event, NorthEast Regional Computing Program, Southbridge, MA.

University of Phoenix launches PhoenixMobile app for iPhone and iPod touch. (2011, April 26). Retrieved from http://www.prnewswire.com/news-releases/university-of-phoenix-launches- phoenixmobile-app-for-iphone-and-ipod-touch-120695069.html

Wentzel, P., van Lammeren, R., Molendijk, M., de Bruin, S., & Wagtendonk, A. (2005). *Using mobile technology to enhance students' educational experience*. Retrieved from EDUCAUSE Center for Applied Research website: http://net.educause.edu/ir/library/pdf/ers0502/ cs/ecs0502.pdf

Wexler, S., Schlenker, B., Brown, J., Metcalf, D., Quinn, C., Thor, E., van Barneveld, A., & Wagner, E. (2007, July). *Mobile learning: What it is, why it matters, and how to incorporate it into your learning strategy. 360° Report, The eLearning Guild*. Retrieved from https://wiki.ucop.

edu/download/attachments/34668692/360report_mobile_complete.pdf?version=1&modifica
tionDate=1275608665000

WirelessWeek.com. (2010a, March 7). eBook reader market to hit $2.5B. *Wireless Week.*
Retrieved from http://www.wirelessweek.com/By-The-Numbers/2010/03/Policy-and-
Industry-BTN-March-2010-Research/

WirelessWeek.com. (2010b, March 7). Global mobile workforce still growing. *Wireless Week.*
Retrieved from http://www.wirelessweek.com/By-The-Numbers/2010/03/Policy-and-
Industry-BTN-March-2010-Research/

Zickuhr, K. (2011, February 3). Generations and their gadgets. Retrieved from http://www.pewin-
ternet.org/Reports/2011/Generations-and-gadgets.aspx

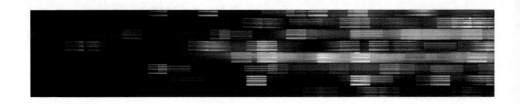

· 6 ·

COLLABORATIVE TECHNOLOGIES

The water cooler of the past has shifted to new virtual venues. We have discussed how people have become less PC-oriented and more mobile-friendly, but another change is also tangentially occurring. Individuals are moving away from being "place-centric"—that is, being focused on meeting people at a given location—to being more "people-centric," or having a focus on connecting with each other online and in virtual space. People increasingly expect to be able to engage with others regardless of location and geographic boundaries.

When technology-based platform concepts were first introduced in the late 1980s, experts predicted that not only would it make a difference in productivity in the future, but the very definition of an office would be transformed (Cox, 2011). At workplaces today, technology-enabled collaborations have been made possible by uniting communications and computing technologies, such as voice mail, instant messaging, chat forums, blogs, microblogging, wikis, social networking sites, voice over Internet protocol (VoIP) telephony, and video conferencing and telepresence. Collaborative technologies provide a unique virtual platform for an unprecedented range of options and media, from personal expression via the sharing of ideas, information, videos, links, images, and audio content, to global business management and information sharing through telepresence, virtual teams, and cloud computing.

> As more people interconnect online, we increase our capacity for both independence and interdependence. Competition and cooperation both thrive in our new culture. The global Internet fosters numberless combinations of groups of every size, sponsoring mass individuality and massive participation. Cyberspace is a vast new civilization, containing both places of commerce and an already deep social life mirrored in countless conversations.
>
> —Jessica Lipnack and Jeffrey Stamps, *Virtual Teams: People Working across Boundaries with Technology*

While collaborative technologies are creating a highly networked world, often supplanting face-to-face interactions with digital alternatives, issues of "trust" and the individual, in relation to work, professional networking, and education, have arisen. Trust, privacy, and security have also become important Internet and social networking concerns. To what extent does trust factor into how new technologies impact peoples' relationships, day-to-day work environments, and educational settings? How should higher education institutions prepare for peoples' expectation and adoption of these technologies for learning? To answer these questions, we begin with a brief overview of some of the popular tools in the modern-day communication toolbox.

Tools of the Trade

The tools commonly used today for collaborating across time zones, cultures, and geographical boundaries have gone far beyond the familiar options of voicemail, email, and fax. Information is now exchanged using a combination of Web applications, mobile devices, and the Internet commonly referred to as Web 2.0: wikis, blogs, online publishing, Internet forums and discussion boards, instant messaging, online chats, and Internet-based TV and telephones. We also communicate through social networking sites (e.g., Facebook, Twitter, Myspace, LinkedIn, and YouTube); virtual environments (e.g., Web conferencing and telepresence); and cloud computing, in which remote servers execute computing, software management, and Web-based functions to provide on-demand and potentially limitless application and storage possibilities for individuals and companies.

Web 2.0: Growth and Adoption

The term *Web 2.0* was coined by Tim O'Reilly, CEO of the computer book publishing company that bears his name, and Media Live International as they reassessed the state of the computer industry in the wake of the dot-com bust during the autumn of 2001. At the time, Web 2.0 was meant to explain the shift from the relatively static nature of existing Internet applications and information-gathering activities to the next leap in communication to a more dynamic and interactive Internet, where information is gathered quickly and readily accessible. The term has since evolved into a marketing buzzword to encompass a broad range of commonly used Web applications and services, such as wikis, blogs, podcasting, and the like (O'Reilly, 2005).

To offer a brief recap of these popular tools: A *wiki* is a website that allows users to create and edit the content of interlinked Web pages using a software program and a Web browser. Computer programmer Ward Cunningham developed the first wiki in 1994. Its technology of Web-based open-source software was later adopted by Jimmy Wales and Larry Sanger in 2001 to create Wikipedia, an online encyclopedia, which today includes 15 million freely usable articles on a vast array of topics that people all over the world can contribute to and edit ("History of Wikipedia," n.d.). *Blogs* are websites that allow information to be exchanged on any given topic as part of an ongoing dialogue. Online chat and instant messaging involve exchanging short text communications in real time through Web pages, dedicated programs, or apps on computers or cell phones. *VoIP* is another popular tool that uses the Internet to transmit voice, allowing people to make phone calls online. Podcasting, as described in the previous chapter, allows the listener or viewer to use an application known as a "podcatcher" to access Web feeds, check for updates, and download new files ("Podcast," n.d.). Blogs, wikis, podcasting, and chatting have created a virtual platform of countless, instant discussion boards and forums that are freely available on the Web.

Social Networking

While social networking sites are considered part of Web 2.0, they have acquired a unique status for their profound influence on society today. *Social networking* is the use of Internet and computing technology to create real-time, user-generated Web communities. By creating profiles on a social network's platform to store and distribute select information about themselves, users can link up with existing friends or other individuals with similar backgrounds or interests and share information, media, or links to third-party online content. Purposes for social networking include establishing or maintaining relationships (Facebook, Myspace), business research and job hunting (LinkedIn), forming special-interest groups or pursuing citizen activism (Meetup.com), sending updates about one's activities or location (Twitter, Foursquare), and enterprise collaboration (Yammer, Chatter).

One of the best-known social networking sites is Facebook. Launched in 2004, Facebook allows online users from around the world to post photos and create online profiles. They can send public and private messages, join interest groups, contribute to the pages of people or organizations and indicate their "likes" for them, or chat online. In 2010 Facebook had more than 500 million active users, with people spending over 700 billion minutes per month on the site; around 200 million of those users connected through mobile devices. There were more than 900 billion pages, groups, events, and so on for users to interact with, and the average user connected to about 80. Facebook is a truly global phenomenon, and is available in 70 languages, with 70% of its users located outside of the U.S. (Facebook, 2010). By the third quarter of 2010 the growth in Facebook users had leveled off and had even dropped by 10 million (Kiser, 2010).

Like Facebook, Twitter is another popular social networking site that specializes in microblogging, or posting text-based messages of no more than 140 characters on profile pages. A posted message is called a *tweet*. One user can subscribe to the tweets of another, which is called "following" that person or organization. Users' effectiveness can be determined by their number of followers. Messages can be received on smart phones and other mobile devices. Launched in 2006, Twitter hosted 65 million tweets a day by 2010.

LinkedIn is a social networking site used primarily by people seeking work or making business contacts. The site is also used for posting resumes and for other kinds of professional self-promotion. Users of LinkedIn tend to spend their efforts online attempting to convey what they know or what they can do; the site is comparatively less interactive than other social networks. This may explain why LinkedIn has significantly fewer users than sites like Facebook and Twitter (Kiser, 2010). Nevertheless, LinkedIn's global reach is roughly 100 million users across 200 countries, with more than half of its users currently located outside of the U.S. ("About Us," 2011).

Yammer is secure online social networking software, specifically designed for internal corporate or business communication. Employees use it for microblogging, setting up profiles, group discussions, direct messaging or sharing files, links, and messages. It also features other capabilities such as a company directory, knowledge base, and downloadable applications for mobile devices. Chatter is another business social networking tool for connecting with co-workers and managing projects, people, and profiles.

Video Technology

The growth of video technology for mass use on the Internet has fueled the unprecedented popularity of YouTube, the leading online video sharing website. Users upload short videos and make them available to everyone online. Users who are not registered with the site can watch all the videos, but registered users can also upload them if they choose. There are videos on how to make things or do things, like build a cat fence or a garden composter. YouTube was launched in 2005 and became an instant hit. In May 2010 alone, 14.6 billion videos were viewed on YouTube. In the U.S., 178 million U.S. users watched some kind of YouTube video in April of 2010; in just one month, that total rose to 183 million (Ehrlich, 2010). Video technology such as YouTube that promotes face-to-face, visual content that people can produce and share anytime and anywhere continues to be more popular than ever. Online digital-data backup companies anticipate steep demand for digital storage in the future, particularly as complex graphics, videos, 3D animation, and smart phone use become more widespread (XZ Backup, 2011).

Business Insight (2011) predicts that services that stream content over the Internet will be in high demand in coming years. With the rise of Netflix and Hulu, which provide TV content on demand, the concept of downloading content is becoming less attractive compared to accessing it at will when needed. Streaming video-rental services, such as iTunes and Netflix, and streaming video game services such as OnLive that allow users to rent video games rather than buy them, allow for greater consumer flexibility and ease of use.

Video Conferencing and Telepresence

Video conferencing generally refers to any communications technology that uses both audio and video to connect people who are in remote locations, as if they were meeting in the same room. At its most basic level this technology requires each user to have a computer, Webcam, microphone, and a broadband Internet connection (Anissimov, 2010).

Telepresence represents a combination of technologies that allows its users to feel as if they were in the same room together, with very high-quality visual, auditory, and even tactile stimuli. It takes many different forms, from enhancing the experience of a televised audiovisual connection, to something more akin to virtual reality. A popular application is video-teleconferencing, a higher level of traditional videoconferencing providing improved audio and video quality. It is considered superior to phone conferencing because the visual aspect greatly enhances communication, allowing for real-time perception of facial expressions and body language ("Telepresence," n.d.).

There is a difference between video conferencing, which has been around for a longer time, and the newer telepresence technologies that make the participants feel that they are actually in the same room with each other. Forrester Research says that the goal of telepresence technology is to make everyone involved in the meetings feel as if they are actually in the room with the other people attending, even when the attendees are actually in several different physical locations (Shein, 2010).

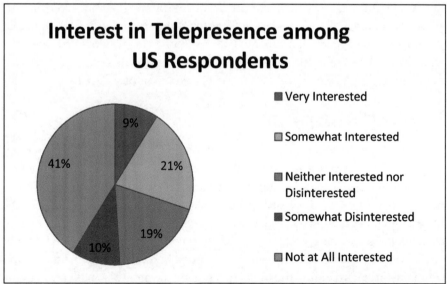

Source: "George Jetson, Meet Cisco's Telepresence," by B. Piper, 2010, retrieved from http://multiplayblog.com/2010/10/06/george-jetson-meet-ciscorsquos-%C5%ABmi-telepresence.aspx

Cloud Computing

Cloud computing is a metaphor for the use of remote server networks, rather than an individual personal computer, to store software applications and access data, information, or computing power (Mell & Grance, 2009). Cloud service providers offer businesses and individuals on-demand software, networking, computing platforms, and infrastructure without requiring on-site maintenance or user programming. Mobile wireless devices are increasingly becoming the means by which users communicate, store and exchange data, and access applications in the cloud (Anderson & Rainie, 2010).

The unique feature of this technology is that users are not limited by their computers' hard drives. With cloud computing, they can access their data anywhere and anytime through an Internet link. Similarly, companies will have the option of investing less in costly hardware and using "the cloud" for applications and storage. Salesforce.com is a popular cloud computing service, providing companies a way to offload traditional functions as managing databases, customer service, and troubleshooting.

The most popular cloud computing uses are for social networking (including the 500 million users of Facebook), Webmail (Hotmail, Yahoo, and Gmail), Web-based blogging and messaging (Twitter, Tumblr, and WordPress), media sharing (YouTube, Flickr), document collaboration (Google Docs), social bookmarking (Delicious), business (eBay, Amazon Web Service), and ranking/rating/commenting (Yelp, TripAdvisor; Anderson & Rainie, 2010).

Social Collaboration

A 2010 survey showed that Americans spend an average of 2.7 hours per day accessing the Internet through mobile devices, and that 91% of those mobile users go online to use social networking sites. Compare this with the 79% of non-mobile users who use social networking (WirelessWeek.com, 2010). Internet users in the U.S. spend an average of one quarter of their time online using social networking sites, blogs, and other collaborative tools—an increase of 15.8% from 2009. All told, Americans spend one third of their time online communicating, including the use of social networking sites, email, and instant messaging (NielsenWire, 2010)

Who Are the Users?

Social networking sites consume a great deal of our time and attention, as the statistics bear out:

- In 2011, Facebook had 500 million users, with 1 in every 13 people on earth a Facebook user. The U.S. share is 206.2 million, which means 71% of the Web audience online is on Facebook. Two hundred fifty million users log in every day, and the average user has about 130 friends. One out of three users are 35 years old or over, with the 18–24 user group seeing the fastest growth in one year. Statistics also show that over 700 billion minutes a month are spent on Facebook, 20 million applications are installed per day, and over 250 million people interact with Facebook from outside the official Website on a monthly basis, across 2 million third-party Websites. Over 200 million people access Facebook via their mobile phone, and 48% of young people said they now get their news through the service. Meanwhile, in just 20 minutes on Facebook over 1 million links are shared, 2 million friend requests are accepted, and almost 3 million messages are sent (Digital Buzz, 2011).

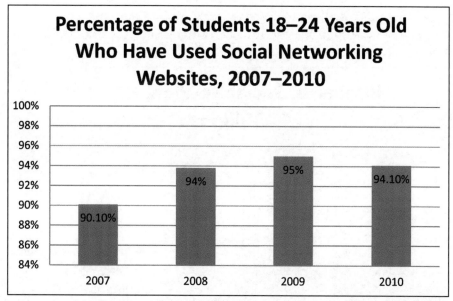

Percentage of Students 18–24 Years Old Who Have Used Social Networking Websites, 2007–2010

Source: "Week 10: Information Technology and Education," by P. Carr, 2011, retrieved from https://impactofinformationsystemsonsociety.wordpress.com/2011/03/09/week-10-information-technology-and-education/

- Twitter users also tend to be older than might be expected. In the third quarter of 2010 only 4% of Twitter users were 17 years or under; 13% were 18–24, 30% were 25–34, 27% from 35–44, 17% from 45–54, 7% from 55–64, and a tiny 2% were 65 or over (Kiser, 2010). Twitter users are younger than Facebook users, but the largest numbers are adults between 25 and 45, not students or younger adults. Internet users in lower-income households are somewhat more likely to use Twitter. Around 17% of those in households earning less than $30,000 per year updated their status with messages online, while only 10% of users in households with an income of over $75,000 per year did so. This may be because the people in lower-income households tend to be younger.

- Mobile users who access the Internet through laptops, cell phones, and handheld devices are more likely to tweet. Approximately 14% of mobile users tweet, while only 6% of non-mobile users do. They are also more likely to get news through their mobile devices. People who use other social networks and who blog, or post articles and comments online, are also more likely to use Twitter (Lenhart & Fox, 2009).

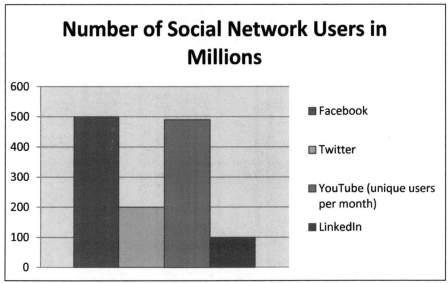

Sources: "Twitter Hits Nearly 200M Accounts, 110M Tweets Per Day, Focuses On Global Expansion," by O. Chiang, 2011, retrieved from http://www.forbes.com/sites/oliver chiang/2011/01/19/twitter-hits-nearly-200m-users-110m-tweets-per-day-focuses-on-global-expansion/; "10 Fascinating YouTube Facts That May Surprise You," by A. Elliott, 2011, retrieved from http://mashable.com/2011/02/19/youtube-facts/; "100 Million Members and Counting...," by J. Weiner, 2011, retrieved from http://blog.linkedin.com/2011/03/22/linkedin-100-million/

- LinkedIn users also tend to be somewhat older than users of other social networking sites. A mere 1% of them are under 17, which is not surprising, given the purpose of the site. Only 3% of young people from 18–24 use the site, which may be because they lack the educational background and training usually found in a detailed LinkedIn entry. Out of LinkedIn's user base, 20% are 25–34 years old, 35% are 35–44, 25% are 45–54, 12% are 55–64, and retirement-age users 65 or over make up only 3%. Men comprise 57% of LinkedIn's users (Kiser, 2010).

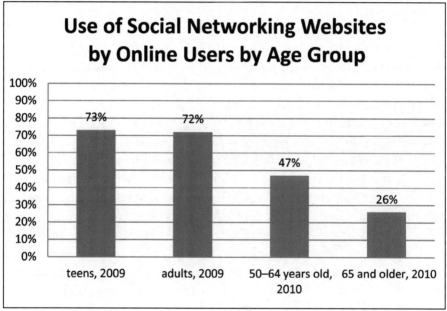

Source: "Older Adults and Social Media," by M. Madden, 2010, retrieved from http://www.pewinternet.org/Reports/2010/Older-Adults-and-Social-Media/Report.aspx

What Is the Impact?

The kind of instant, on-the-move communication that reaches millions every day is bound to have a social impact. The 2008 presidential election, for instance, showed some of the potential that these tools can have. According to *The New York Times*, Barack Obama's presidential campaign changed American politics forever by introducing a new way to campaign that used the Internet and social networking technology in particular. The Obama campaign employed Web 2.0 tools to organize supporters, advertise, communicate, and answer attacks, all without going through the intermediary of the press. The Internet and social technology also allowed voters to check claims made by the candidates and share information about them instantly (Miller, 2008).

Among military families, social media has become increasingly important as a way to keep in touch across continents. A survey of over 3,500 families by Blue Star Families, a nonprofit organization created by military families, showed that nearly 90% of those responding used some type of social media. Eighty-eight percent said they used social media at least once a week. During deployment, the families of military personnel rely heavily on social media, with 89% saying that they use email to communicate with those sent overseas (Blue Star Families, 2010).

Social collaboration technologies are being used in ad hoc ways for crisis management and humanitarian efforts (O'Reilly & Battelle, 2009). During the 2010 Haitian earthquake, for instance, a Kenyan-born organization called Ushahidi deployed a crisis-mapping site, which allowed users to submit eyewitness accounts and other relevant information via email, text, and Twitter. These data points were then mapped to allow the distribution and frequency of events to be visualized (Tapscott & Williams, 2010, p. 4).

In 2011, the global reach of social media sites like Facebook, Twitter, and YouTube has been associated with political upheaval in the Middle East and Africa. A young generation of users in Tunisia and Egypt, for instance, used the real-time, instant feedback features of social networking technologies to bring together masses of people for large-scale protests and rallies. In another famous incident, the killing of Osama bin Laden in 2011 was live-tweeted by someone who was not aware that the event was actually taking place, but later realized the magnitude of his tweets as events unfolded on the international scene (O'Dell, 2011).

In many ways social networking and Web 2.0 technologies are allowing people to act at a more grassroots level, cutting out the intermediary. Celebrities are tweeting their fans directly, giving the fans glimpses into their private lives, rather than relying on public relations experts to relay information to the media. Padmasree Warrior, Chief Technology Officer of Cisco Systems, is said to have 1.4 million followers on Twitter, which she dubs a "digital water cooler" of sorts that provides live feedback to help shape one's thinking ("Padmasree Warrior," 2011). News is being broadcast immediately to millions, both through individual efforts and through media representatives that use social networking sites to promote their services. Goods and services are being marketed directly, and unusual talent is being discovered on viral videos that strike a chord with the online public.

Telepresence and video-based technologies are also profoundly altering how we conduct business and engage with co-workers worldwide. People are already talking to one another all over the world using Webcams and microphones. Network World has reported on a new telepresence phenomenon called *tele-flirtation*. Some telepresence and videoconferencing technology is so real that strangers are connecting through telepresence meetings over the Internet (Network World, 2010). In another innovative development, 3D holographic display technology maker Musion Systems (musion.co.uk) and Cisco (cisco.com) are partnering to create the Cisco TelePresence On-Stage Experience, the world's first real-time virtual presentation.

What Lies Ahead?

Social networking sites are popular today, but explosive growth in the U.S. is expected to plateau in 2012. Experts believe that by then, more than half of all Internet users will have been using social networking sites, and fewer will be expected to sign up. The next step, according to forecast analysts, will be to keep users engaged via new and meaningful content and experiences (webpronews.com).

A survey by the Pew Research Center's Internet & American Life Project and Elon University's Imagining the Internet Center states that by 2020, most people will work on cloud-based applications and software in place of software running on a general-purpose PC (Anderson & Rainie, 2010). The report states:

> Some experts in this [Pew] survey said that for many individuals the switch to mostly cloud-based work has already occurred, especially through the use of browsers and social networking applications. They point out that many people today are primarily using smartphones, laptops, and desktop computers to network with remote servers and carry out tasks such as working in Google Docs, following Web-based RSS (really simple syndication) feeds, uploading photos to Flickr and videos to YouTube, doing remote banking, buying, selling and rating items at Amazon.com, visiting with friends on Facebook, updating their Twitter accounts and blogging on WordPress.

Desktop use, however, is not in decline yet. According to the Pew report, desktops will be used in tandem with remote computing, and users will employ a desktop/mobile hybrid, whose form will eventually evolve in response to users' needs. Moreover, acceptance of cloud computing via mobile devices also will depend on the cost, availability, and reliability of broadband Internet access across the world (Anderson, 2010).

Since social networking sites, telepresence, and video-based technologies have been around for less than seven years, there can be no doubt that we have only seen the tip of the iceberg in the way this instant access to millions of individuals and organizations is going to change our ways of connecting socially and addressing issues and events that concern us—indeed, our way of life.

Collaborative Technologies in the Workplace

Web 2.0 and social networking sites have taken some time in gaining acceptance in the workplace. For instance, employers often discourage use of sites like Facebook during work time, since they tend to distract the workers. Similarly, there are frequent headlines about workers who have been fired for posting inappropriate information on a social networking site. In recent years, however, many organizations are beginning to recognize the potential benefit of including social networking technology in their approach to doing business. Social networking is seen as a way for businesses to establish trust and develop a relationship with their customer base. It can also enable them to find qualified workers and allow communication among existing workers.

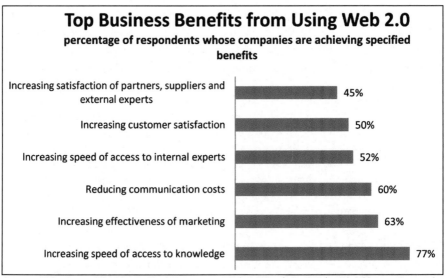

Top Business Benefits from Using Web 2.0
percentage of respondents whose companies are achieving specified benefits

Benefit	Percentage
Increasing satisfaction of partners, suppliers and external experts	45%
Increasing customer satisfaction	50%
Increasing speed of access to internal experts	52%
Reducing communication costs	60%
Increasing effectiveness of marketing	63%
Increasing speed of access to knowledge	77%

Source: "The Rise of the Networked Enterprise: Web 2.0 Finds Its Payday," McKinsey Quarterly, 2011, retrieved from https://www.mckinseyquarterly.com/Organization/Strategic_Organization/ The_rise_of_the_networked_enterprise_Web_20_finds_its_payday_2716?pagenum=2

> People need other people to realize their greatest impact, and innovation, perhaps the most valuable activity in business, depends critically on the kind of cross-pollination of ideas that collaboration enables.
> —Jeffrey F. Rapport, *Technology Review,* March 2011

In addition to greater acceptance and use of social networking, companies are also utilizing the benefits of video conferencing and telepresence technologies, for a number of reasons: They reduce the cost, time, and resources associated with travel, and they help businesses reduce their carbon footprint by using less fossil fuel to travel around the globe.

Small Business Use

Businesses have found that social networking sites can be effective marketing tools for word-of-mouth advertising. They put "Tell a Friend" tools on their websites and then encourage visitors to share this information with friends in their social networks. Facebook is the most popular venue for this kind of advertising, with Myspace second, and Twitter third. Twitter also sells tweet-sized advertisements for $100,000 per day, but the response rate on these ads is so high that it is worth the price. This is increasing Twitter's share of advertising revenue and is once again changing online promotion (Carr, 2010).

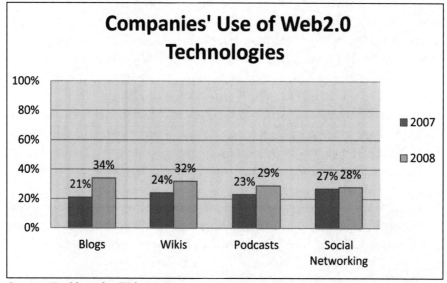

Source: "Building the Web 2.0 Enterprise: McKinsey Global Survey Results," McKinsey Quarterly, 2011, retrieved from https://www.mckinseyquarterly.com/Business_Technology/BT_Strategy/Building_the_Web_20_Enterprise_McKinsey_Global_Survey_2174?pagenum=2

In a 2011 *USA Today* business article, several trends affecting small businesses in the coming years were analyzed. All are technology-related and are bound to impact the way businesses connect with customers. They include mob discounts (e.g., online coupon services such as Groupon.com that are linked to social networking sites); mobile apps to locate deals depending on the users' location; and having a Facebook presence, which establishes the reach and brand of a company (Strauss, 2011).

Corporate Use

While the idea of collaboration among workers has always had acceptance, in modern terms it means working together using technology and social networking to connect people, ideas, and organizations across geographic boundaries. Cisco refers to this activity as *Collaboration 2.0*, where the focus is a shift from documents and PC to people and "social sessions." The goal of social sessions, according to Cisco, is "to create the effect of presence within the reality of absence." In other words, with the help of wikis, blogs, social networks, and a variety of conferencing devices, organizations can find experts in an instant, which makes for a better business environment (Cisco, 2010).

Cisco's use of telepresence is an example of how collaboration can benefit workplace productivity, such as cutting business expenses by reducing the need to travel to meetings and allowing the review of projects requiring visual quality control (Engbretson, 2010). Cisco is the main participant in the telepresence market. Other technologies in this area are Polycom's UC Everywhere™ (Polycom, 2011), HP's Halo™ and HD videoconferencing platforms (HP, 2010), and Vidyo's videoconferencing capabilities for mobile software platforms, which allows users of smartphones and tablets to meet virtually with colleagues who are using PCs and full-room conference systems (Lawson, 2010). Another technology of interest is the use of robots that attend meetings. Anybot telepresence robots are armless human-sized bots on wheels that let companies remotely view and interact with people, as well as exchange files (Saenz, 2010).

Companies are finding many uses for telepresence technology. The insurance giant MetLife uses it primarily for internal meetings and intra-company collaborations. MetLife's vice president for employee benefits sales says:

> Telepresence allows me to see and virtually interact with more people on my team, instead of just my direct reports. When we use telepresence for meetings, people who wouldn't normally be asked to travel to New York have the opportunity to make pre-

sentations and get valuable exposure to executive management. It really facilitates face-to-face interaction with a broader cross-section of employees on an economically efficient basis. (Shein, 2010, p. 22)

The law firm of Lathrop & Gage, LLP, in Kansas City, Missouri uses videoconferencing and more high-end telepresence equipment to hold around 300 meetings a month. The attorneys find these virtual meetings more useful and beneficial than the telephone conferences that they have replaced. The firm's CIO, Ben Weinberger, estimates that an attorney can save $1,500 in travel expenses and productivity by attending one virtual meeting instead of traveling to attend in person. Since law is a billable-hour profession, this saving can help a firm cut costs and give it a competitive edge. Weinberger estimates that a single attorney could save $30,000 a year by using this technology (Shein, 2010).

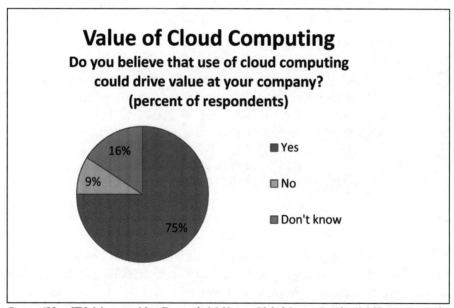

Source: "How IT Is Managing New Demands: McKinsey Global Survey Results," McKinsey Quarterly, 2011, retrieved from https://www.mckinseyquarterly.com/Business_Technology/BT_Strategy/How_IT_is_managing_new_demands_McKinsey_Global_Survey_results_2702?pagenum=3

Telepresence technology has transformed other aspects of business besides meetings, training, and product-development collaboration. It can facilitate visual approval for industries engaged in design or that rely upon quality assurance. One advertising agency, for example, now uses it to review mockups with remote clients. In the past, these had to be sent by overnight delivery before they could be discussed. Now, discussion, review, and revisions can happen in real time. A clothing manufacturer is also using telepresence technology to review and approve samples being manufactured offshore (Engebretson, 2010).

A 2011 Deloitte report predicts that cloud computing will become a collaboration technology of choice as businesses increasingly engage its capabilities for such activities as managing applications and data storage, as well as sharing critical information. Deloitte points to the healthcare industry as a prime candidate for this technology as it moves several functions to the cloud, such as recruiting, background checks, scheduling interviews, and bringing new talent on board (Deloitte, 2011).

Interestingly, as businesses continue to promote new concepts and technological changes in the workplace, there are some traditional collaborative tools that remain a strong first choice among workers. A 2010 Gartner survey of 416 U.S. companies showed, for instance, that email was the most popular tool among workers, followed by group calendaring/scheduling, Web conferencing, team workspace, and other miscellaneous work-related management tools. The survey also noted, however, that investment in social networking technologies is on an upward trajectory as companies continually prioritize software needs based on employee use and preference as well as the value added to the enterprise (Cox, 2011).

Impact on Higher Education

Much like the business world, educational institutions are coming to terms with the advantages and disadvantages of social networking sites. In many grade and secondary schools, sites like Facebook and Twitter remain blocked, and schools mainly become concerned about social networking when debating whether they should punish students for posting inappropriate material on the sites, engaging in cyber-bullying, and so on. However, a Pew Research study released in 2010 showed that 73% of American kids age 12 to 17 uses some kind of social networking site, up from 55% in 2006. Some educators have concluded from these data that, since social networking clearly is here to say, it would be better to exploit its potential in the educational setting and teach students how to use it responsibly (Davis, 2010). In higher education institutions, social networking and collaboration technologies are the new reality as the 18-and-over demographic uses them on an almost daily basis.

Student and Instructor Use

The preponderance of collaborative technology use in students' personal lives seems to spill effortlessly into the higher education environment. Anderson (2007) aptly summarized Web 2.0 use by educational institutions in four preliminary areas, namely, learning and teaching, scholarly research, academic publishing, and libraries. More than 9 out of 10 student respondents to an EDUCAUSE survey showed that they used text messaging and accessed social networking websites daily, and 4 out of 10 used computer-based VoIP as a replacement for the telephone. Social networking sites most commonly used include Facebook (96%), followed by Myspace. Activities included staying in touch with friends (96%) and sharing photos, music, video, and the like (72%). Web-based technologies commonly used by students include word processors; spreadsheets; presentation and form application tools from Google Docs, iWork, Microsoft Office Live Workspace, Zoho, and so forth; wikis; social networking, video sharing websites, Web-based calendars, and blogs. Here are some interesting ways that Web 2.0 is being incorporated into learning modules:

- Podcasting: With lectures available as podcasts, students can review information at their own pace. This is ideal for students who want the flexibility to listen to lectures when they want, as opposed to going to class at an assigned time. It is also useful for international students who might find language a barrier.

- Blogs: The use of blogs by experts in various fields to share their knowledge is rapidly increasing. For example, Larry Lessig (2008), Harvard University and founder of the Center for Internet and Society while at Stanford Law School, engaged participants in an active dialogue about copyright issues and the Internet. Because blogs encourage commentary, students learn not simply by absorbing information, but also by processing, analyzing, and responding to it.

- Wikis: Wikis provide an ideal medium for students to pool their knowledge and learn by teaching one another. Using wikis, students can comment on information, spurring analysis and discussion. Wikis can also incorporate multimedia such as animation, voice, music, and video.

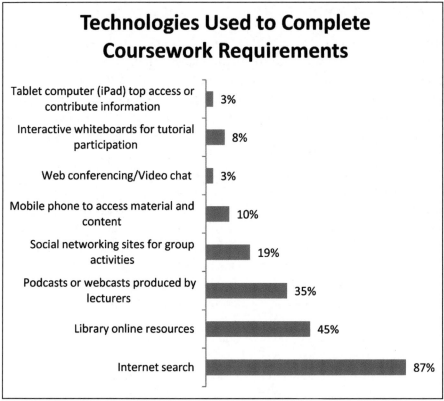

Technologies Used to Complete Coursework Requirements

- Tablet computer (iPad) top access or contribute information — 3%
- Interactive whiteboards for tutorial participation — 8%
- Web conferencing/Video chat — 3%
- Mobile phone to access material and content — 10%
- Social networking sites for group activities — 19%
- Podcasts or webcasts produced by lecturers — 35%
- Library online resources — 45%
- Internet search — 87%

Source: "Reputation Management and Social Media, Part 2: Concerns about the Availability of Personal Information," by M. Madden and A. Smith, 2010, retrieved from http://www.pew internet.org/Reports/2010/Reputation-Management/Part-2/Managing-identity-through-social-media.aspx

Another interesting trend in recent years has been the steady growth of online tutoring sites, where individuals or groups of students use technologies such as Skype, YouTube, or online videos to get instruction and help from tutors in locations around the world. The Khan Academy (khanacademy.org), for example, provides over 2,000 custom self-paced learning videos free of cost for students to watch or download to supplement their learning.

Educational institutions have found creative uses for telepresence technology in student instruction. Barrow Neurological Institute uses it to allow students to view brain surgery while it is taking place. By using this technology, the students do not disturb the surgeon or take up space in an operating room, and all of them have the same clear view of the procedure. Duke University uses telepresence gear to present classes in other countries, which expands the university's revenue base. It also allows the overseas students an educational opportunity they most likely would not be able to afford if they had to travel to attend classes. This equipment might also be used to allow professors to conduct office hours from remote locations (Engebretson, 2010).

While some instructors are leveraging the educational possibilities of social networking and video-based technologies, there is also a general perception that many cling to old ways of doing things. A 2010 EDUCAUSE study, for instance, found that instructor use of information technology resources, including technologies for collaborative purposes, was somewhat different than students' use. While some instructors relied on course management systems for assignments, exams, discussion forums, and so forth, fewer used collaborative editing tools, blogs, plagiarism detection tools, student response systems, or video/game/simulation/virtual worlds for instructional purposes. Despite opportunities to engage in these type of technologies, it was found that instructors preferred to use old-school, lecture-based instruction (Smith & Caruso, 2010).

The study also showed that students have, for the most part, a positive perception of information technology in the classroom, citing its convenience, ability to facilitate learning, and the opportunities it provides to become actively involved in the learning process. Most, however, preferred a moderate use of technology in instruction, suggesting a hybrid approach as opposed to extremes of all-tech or no-tech at all (Smith & Caruso, 2010).

Institutional Use

Colleges and universities are learning that social networking sites help them create an online presence, as well as a channel to promote their schools, keep alumni updated, solicit donations, support athletic teams, and recruit students (Gilroy, 2009). One administrator has used these sites to request technology and equipment donations for a school and then used the project to attract national media attention to the school (Davis, 2010). Teachers at various levels of academia use social networking sites to connect with colleagues and generate ideas about new approaches to education.

Students now find that a university's electronic resources are critical to many of their activities. Many rely on university e-resources, such as library Websites, presentation software, and course management systems (CMS) for communication and learning (Smith & Caruso, 2010). CMS tools, which include software systems such as Blackboard, WebCT, Desire2Learn, Sakai, and others, permit the exchange of course-related information such as presentations, homework, exams, grades, and discussion threads. Students also routinely expect to use their university websites for a variety of services, from course registration and paying tuition to downloading college-approved software programs for learning and instruction.

Telepresence has widespread potential for allowing educational institutions and businesses to expand their services globally. Cisco Systems, one of the leading marketers of telepresence technology, has a pilot program that it hopes to develop into a global educational enterprise. This "global U" would allow professors to teach students from all over the world and allow students to choose from a virtually endless list of classes. All of this could happen without anyone having to travel further than their local telepresence room. By the autumn of 2009 Cisco had signed up around 35 educational institutions, including Duke University's Fuqua School of Business and the University of South Carolina's Moore School of Business.

As student use of technology accelerates, educational institutions are finding that they too must constantly evolve to stay ahead of the game. They understand that they need to be proactive to successfully leverage student use of mobile devices for a host of educational and administrative functions. They are also realizing that it is just as important to make smart choices when investing in these technologies. Identifying and prioritizing resources, and gauging student usage, are all factors that come to bear, for instance, when considering whether to buy information technology tools from outside vendors or to build applications in-house, or even whether to use open-source platforms such as those developed by MIT, which are free to use (Keller, 2011). To ensure that universities allot limited funding on the most crucial and accepted equipment and software, they must remain attuned to the cornucopia of social uses of technology that students bring to campus life.

Privacy and Security Issues

Trust and privacy are important considerations that underlie the use of social networking sites, which by design encourage users to share personal information while connecting with old friends or making new ones. The role of privacy in social networking was examined in a study that compared users of Facebook and Myspace regarding the extent to which individuals were willing to make new friends online (Dwyer, Hiltz, & Passerini, 2007). While members of both sites expressed privacy concerns, the study showed that trust is not as necessary in the building of new relationships online as it is in face-to-face encounters.

Also, the willingness to share information does not automatically translate into new social interactions. One of the reasons may be that users had the option to "pull the plug," or ignore and block messages, thus minimizing the risk of exploring online relationships.

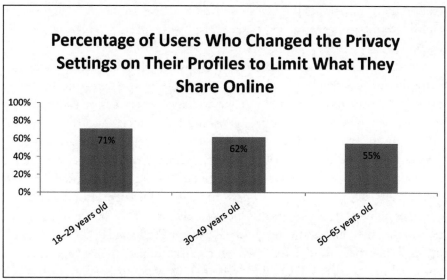

Percentage of Users Who Changed the Privacy Settings on Their Profiles to Limit What They Share Online

Source: "Reputation Management and Social Media, Part 2: Concerns about the Availability of Personal Information," by M. Madden and A. Smith, 2010, retrieved from http://www.pew internet.org/Reports/2010/Reputation-Management/Part-2/Managing-identity-through-social-media.aspx

Although young Internet users freely share information online, they are also far more active than older users in "curating" this information; for instance, 71% of users aged 18–29 changed the privacy settings on their profiles to limit what they share online (Madden & Smith, 2011). This is compared to 62% of users age 30–49 years and just 55% of users age 50–65. The study also shows that young users are not only the most proactive in customizing their privacy settings, but they trust social networking sites less than older users do. Specifically, when how much of the time they can trust these sites, 28% of users age 18–29 say "never," compared to a smaller percentage in other age groups. This seems to suggest that while young users are aware of trust and privacy concerns associated with social networking sites, they are also confident of their ability to use the technology in ways that can mitigate that risk.

Many privacy advocates worry about how our digital footprint is making life less private as the Internet, and particularly social networking, aggregate data about our likes and dislikes with the intent to deliver personalized content and services. Google's search engine and Facebook use sophisticated data mining technology based on user accounts to deliver customized content. The Global Positioning System (GPS) in a smartphone can track its user's geographic location, and it has been reported that the Library of Congress can archive your tweets even if you cancel your Twitter account (National Public Radio, 2010).

In 2010, Facebook was involved in controversy when its site was accused of being lax in securing personal data. In response to this concern, Facebook agreed to tighten its privacy settings, but anxiety over privacy on Facebook and the Internet in general fuels ongoing debate ("Facebook and Privacy Issues," 2010). On the other hand, in May 2011 Facebook cited Google for invasion of privacy with its use of a new tool called Social Circle. In another incident, Sony's PlayStation Network had to temporarily shut down and alert its 77 million users due to a breach of their private personal data. The issue of online privacy becomes more acute as Americans increasingly turn to the Internet and social networking sites for healthcare advice (Reisinger, 2011) and become more willing to discuss their health in an open forum. According to the Pew Internet survey cited in Reisinger's *CIO Insight* report, 11% of adults have followed their friends' personal health experiences or updates via social networking sites.

Identity management is an essential component for the successful development and growth of user-centric Internet 2.0 services (Alpár, Hoepman, & Siljee, 2011). Privacy issues that concern technology are an ongoing concern in both the private and public sectors. Microsoft and Mozilla are among the Web browser companies looking to develop new do-not-track tools that would prevent monitoring of online use. The Federal Trade Commission recently called for such a do-not-track system (Angwin, 2011), and a bill has been introduced in Congress that would give the FTC more authority on these matters. In a 2010 cybersecurity strategy report, the U.S. Department of Homeland Security noted that a more secure cyberspace was critical to the health of the economy and the security of the nation as a whole. In particular, it stated that the federal government must address the recent and alarming rise in online fraud, identity theft, and misuse of information online (*National Strategies*, 2010).

As social networking technologies become more integrated with mobile devices, and Internet use seeps into many of our day-to-day activities, discussions of trust, privacy, and security are bound to intensify, as will the impact of how we choose to use technology in our lives.

Summary

Social networking and collaborative technologies are permeating almost every aspect of our lives today. Their importance grows as people find unique uses for them and bring flexibility, convenience, and enjoyment into their lives. Younger generations are quick to adapt to new Web 2.0 technologies like email, social networking, and video uploading and editing applications, and older adults have also become frequent users. Collaborative and interactive tools such as videoconferencing and telepresence have found use in large corporations as a time-saving, efficient method of doing business. Companies both small and large consider social networking important to their business strategies for improving employee communications and tapping into consumer trends. Collaborative tools are also being adopted in higher education, as institutions tune into the social uses of these tools that students bring to campus. While adoption of high-end tools such as telepresence and video technology is still gaining momentum and acceptance, Web 2.0 applications have become mainstream as instructors use email, blogs, and wikis as essential instruments for communication and content development.

References

Alpár, G., Hoepman, J-H., Siljee, J. (2011, January 2). *The identity crisis: Security, privacy and usability issues in identity management*. Retrieved from Cornell University Library arXiv website: http://arxiv.org/ftp/arxiv/papers/1101/1101.0427.pdf

Anderson, J. Q., & Rainie, L. (2010, June 11). The future of cloud computing. Retrieved from http://pewresearch.org/pubs/1623/future-cloud-computing-technology-experts

Anderson, P. (2007, February). *What is Web 2.0? Ideas, technologies and implications for education*. Retrieved from JISC website: www.jisc.ac.uk/media/documents/techwatch/tsw0701b.pdf

Angwin, J. (2011, February 25). Microsoft endorses do-not-track tool. *The Wall Street Journal*. Retrieved from http://online.wsj.com/article/SB1000142405274870415060457616663281 7863302.html

Blue Star Families. (2010). *2010 Military Family Lifestyle Survey* (Executive summary). Retrieved from http://www.bluestarfam.org/system/storage/42/58/2/301/2010bsfsurveyexecsummary.pdf

Carr, A. (2010, October 11). Twitter crushing Facebook's click-through rate: Report. Retrieved from http://www.fastcompany.com/1694174/twitter-crushing-facebooks-click-through-rate-report

Cox, L. (2011, March 31). Social networks a coming trend at work. *Technology Review*. Retrieved from https://www.technologyreview.com/business/37080/?a=f

Davis, M. R. (2010). Social networking goes to school. *Education Week Digital Directions*. Retrieved from http://www.edweek.org/dd/articles/2010/06/16/03networking.h03.html

Deloitte. (2011). Deloitte analysis of top technology trends for 2011. Retrieved from http://www.deloitte.com/view/en_CN/cn/Pressroom/pr/cb97880465c4e210VgnVCM200000 1b56f00aRCRD.htm

Digital Buzz. (2011, January 18). Facebook statistics, stats & facts for 2011. Retrieved from http://www.digitalbuzzblog.com/facebook-statistics-stats-facts-2011/

Dwyer, C., Hiltz, S. T., & Passerini, K. (2007). *Trust and privacy concern within social networking sites: A comparison of Facebook and MySpace.* Proceedings of the Thirteenth Americas Conference on Information Systems, Keystone, CO. Retrieved from http://csis.pace.edu/ ~dwyer/research/DwyerAMCIS2007.pdf

EDUCAUSE. (2007, April). *7 things you should know about RSS.* Retrieved from http://www. educause.edu/ir/library/pdf/ELI7024.pdf

Ehrlich, B. (2010, June 24). The average YouTube user watched 100 Videos in May. Retrieved from http://mashable.com/2010/06/24/the-average-youtube-user-watched-100-videos-in-may-stats

Engebretson, J. (2010, April). Telepresence not just for meetings anymore: Emerging applications in health care, education and manufacturing offer new opportunities for service providers. *Connected Planet*, 28–30.

Facebook. (2010). Statistics. Retrieved June 22, 2011 from http://www.facebook.com/press/info. php?statistics

Facebook and privacy issues [Editorial]. (2010, October 19). *San Francisco Chronicle.* Retrieved from http://articles.sfgate.com/2010–10–19/opinion/24141616_1_mafia-wars-facebook-privacy-rules

Gilroy, M. (2009). Higher education migrates to YouTube and social networks. *The Hispanic Outlook in Higher Education* Retrieved from http://findarticles.com/p/articles/mi_hb3184 is_20090921/ai_n38021399/

History of Wikipedia. (n.d.). In *Wikipedia.* Retrieved June 22, 2011, from http://en.wikipedia. org/wiki/History_of_Wikipedia

HP. (2010, November 17). HP announces high-definition videoconferencing for desktops and conference rooms [Press release]. Retrieved from http://www.hp.com/hpinfo/newsroom/ press/2010/101117c.html

Keller, J. (2011, January 23). As the web goes mobile, colleges fail to keep up. *The Chronicle of Higher Education.* Retrieved from http://chronicle.com/article/Colleges-Search-for-Their/126016/

Kiser, P. (2010, October). Social media 3Q update: Who uses Facebook, Twitter, LinkedIn, & MySpace? Retrieved from http://socialmediatoday.com/paulkiser/199133/social-media-3q-update-who-uses-facebook-twitter-linkedin-myspace

Lawson, S. (2010). Vidyo brings smartphones into videoconferences. *NetworkWorld.* Retrieved from http://www.networkworld.com/news/2010/052610-vidyo-brings-smartphones-into.html

Lenhart, A., & Fox, S. (2009, February 12). Twitterpated: Mobile Americans increasingly take to tweeting. Retrieved from http://pewresearch.org/pubs/1117/twitter-tweet-users-demographics

Lessig, L. (2008, January 15). *The future of ideas.* Retrieved from http://www.lessig.org/blog/2008/ 01/the_future_of_ideas_is_now_fre_1.html

Madden, M., & Smith, A. (2011, May 26). Reputation management and social media: Part 2.

Concerns about the availability of personal information. Retrieved from http://www.pew internet.org/Reports/2010/Reputation-Management/Part-2/Managing-identity-through-social-media.aspx

Mell, P., & Grance, T. (2009). *The NIST definition of cloud computing* (Version 15, last modified October 7, 2009). Retrieved from National Institute of Standards and Technology website: http://www.nist.gov/itl/cloud/upload/cloud-def-v15.pdf

Miller, C. C. (2008, November 7). How Obama's Internet campaign changed politics. *The New York Times*. Retrieved from http://bits.blogs.nytimes.com/2008/11/07/how-obamas-internet-campaign-changed-politics/

National Public Radio. (2010, May 21). Protecting your privacy on social networking sites. [Transcript of *Science Friday* audio story]. Retrieved from http://www.npr.org/templates/story/story.php?storyId=127037413

National strategies for trusted identities in cyberspace: Creating options for enhanced online security and privacy (Draft). (2010, June 25). Retrieved from U.S. Department of Homeland Security website: http://www.dhs.gov/xlibrary/assets/ns_tic.pdf

NielsenWire. (2010, August 2). What Americans do online: Social media and games dominate activity. Retrieved from http://blog.nielsen.com/nielsenwire/online_mobile/what-americans-do-online-social-media-and-games-dominate-activity/

O'Dell, J. (2011, May 2). One Twitter user reports live from Osama bin Laden raid. Retrieved from http://mashable.com/2011/05/02/live-tweet-bin-laden-raid/

O'Reilly, T. (2005, September 30). What is Web 2.0? Design patterns and business models for the next generation of software. Retrieved from http://oreilly.com/web2/archive/what-is-web-20.html

O' Reilly, T. & Batelle, J. (2009). *Web squared: Web 2.0 five years on*. Retrieved from O'Reilly website: http://assets.en.oreilly.com/1/event/28/web2009_websquared-whitepaper.pdf

Padmasree Warrior joins Lateline Business. (2011, March 3). *Lateline Business*. Retrieved from http://www.abc.net.au/lateline/business/items/201103/s3178229.htm

Podcast. (n.d.). In *Wikipedia*. Retrieved June 21, 2011, from http://en.wikipedia.org/wiki/Podcast

Polycom. (2011). UC Everywhere. Retrieved from http://www.polycom.com/company/uc_everywhere/index.html?id=hpsolhilite&link=uc-everywhere

Reisinger, D. (2011, May 13). Social medicine: Is the Internet transforming healthcare? *CIO Insight*. Retrieved from http://www.cioinsight.com/c/a/Health-Care/Social-Medicine-Is-the-Internet-Transforming-Healthcare-752089/

Saenz, A. (2010, January 25). Anybots telepresence robots go into mass production. *Telepresence Options*. Retrieved from http://www.telepresenceoptions.com/2010/01/anybots_telepresence_robots_go/

Shein, E. (2010, April). Face to virtual face, telepresence technology can slash travel costs—if you can afford it, and if it's really used. *Computer World*, 19–23.

Smith, S. D., & Caruso, J. B. (2010). *The ECAR study of undergraduate students and information technology, 2010* (Research Study, Vol. 6). Retrieved from EDUCAUSE Center for Applied Research website: http://www.educause.edu/Resources/ECARStudyofUndergraduateStuden/217333

Strauss, S. (2011, January). Top 5 trends for small business as 2011 starts. *USA Today*. Retrieved from http://www.usatoday.com/money/smallbusiness/columnist/strauss/2011-01-03-top-5-small-business-trends_N.htm

Tapscott, D., & Williams, A. D. (2010). *Macrowikinomics: Rebooting business and the world*. New York: Portfolio.

Telepresence. (n.d.). In *Wikipedia*. Retrieved June 22, 2011, from http://en.wikipedia .org/wiki/Telepresence

WirelessWeek.com. (2010). Mobile users go online to socialize. *Wireless Week*. Retrieved from http://www.wirelessweek.com/By-The-Numbers/2010/03/Policy-and-Industry-BTN-March-2010-Research/

XZ Backup (2011, April 14). Data and storage growth trends—and how they affect online backup. Retrieved from http://www.xzbackup.com/blog/company-news/data-and-storage-growth-trends-and-how-they-affect-online-backup/

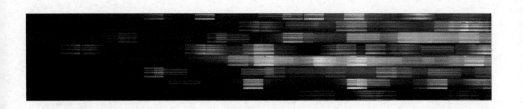

· 7 ·

IMMERSION, GAMING,
AND ROBOTICS

For many people, their introduction to immersive technologies may have
been that first visit to the local IMAX theatre, where everyone wore special glasses to watch images seem to jump off the big screen, endowing viewers with a sense of being part of the action. Today, the IMAX 3D experience
has evolved considerably, and we have the option of not only being a part of
the action, but also being able to respond, communicate, and interact with digital objects and environments in meaningful ways.

The movie *Avatar* is considered a tipping point in the U.S. for transforming 3D technology into a mainstream expectation for U.S. audiences ("The
Champions of 3D," 2011). This feeling of *immersion*, or the blending of artificial and real environments, is garnering attention as a powerful tool for
enhancing many areas of our daily lives, whether it's shopping online for new
clothes or walking through ancient cultures in museums as a way to enrich our
understanding of distant times and places.

Immersion technology is also being actively employed in the world of
gaming, where participants create *avatars*, or digital alter egos that represent
their virtual characters, to engage in 3D virtual environments. Over the period
of a human generation, gaming has evolved from a feature of mall arcades and
something played on cumbersome game systems and early home computers, to
a sophisticated immersion experience taking place in virtual worlds.

Internet gaming today is going mobile on cell phones and other portable devices (Eskelsen, Marcus, & Ferree, 2009). Rather than using games as an opportunity to sit down face-to-face to challenge a single opponent, members of this new generation of game players are engaging with thousands of others online in avatar form.

As immersion and gaming technologies make inroads into our everyday lives, robotics technology is also figuring more prominently in our work world, complementing many human tasks with greater safety and speed. While the ability of machines to do human work has always been the lore of science fiction, real-world applications are unfolding in robotics-assisted surgery, deep-sea oil exploration, and search-and-rescue missions during natural disasters.

Rather than being considered futuristic, robotics, as well as immersive environments and online gaming are the emerging technologies of today and are actively being pursued and researched for uses in education, the military, medical research and practice, and the aerospace, automotive, construction, and manufacturing industries, as well as the corporate and business worlds for teaching, training, and management.

A brief overview of these technologies will show their impact on society, work, and education, and how virtual worlds, gaming, and robotics are imperceptibly informing our lives in new and different ways.

Immersion Technologies

Immersion is the state of being in which users are so absorbed by events and interactions within a virtual world that their consciousness of the physical world is diminished, and one's awareness of the physical self is lost amid a total, often artificial environment ("Immersion [virtual reality]," n.d.). This experience is also referred to as *virtual reality*. Users can immerse themselves through computer-generated digital images and worlds. These virtual worlds are synchronous, persistent networks of people, represented by avatars and facilitated by computers. They are increasingly important to adults and children alike, and can influence how they buy, work, and learn (Bell, 2008).

The growth and development of virtual worlds are continually evolving as new technologies and platforms are created and researchers find innovative ways to integrate their capabilities for enhanced visual images and experiences. Some immersion technologies gaining traction among users include 3D, virtual systems, and augmented reality.

3D Technology

3D technology can be traced as far back as 1838, when Sir Charles Wheatstone first described a phenomenon called *stereopsis*, or the process of overlapping two identical images to create a three-dimensional effect for the human eyes (Sniderman, 2011). Today, 3D technology is being deployed in the movie, video, and TV industries in creative ways.

3D TV has especially gained some interest in recent years among viewers. Two images are presented on the screen that show objects from slightly different angles, and when viewers use special glasses for viewing the screen, the images combine, offering an illusion of depth (Katzmaier, 2010). Many companies are still refining the technology behind 3D TV, and the cost is still high for the average consumer. Some barriers to widespread adoption of 3D TV include whether viewers will actually don the glasses required for viewing on a regular basis, and the expense associated with the shutter glass for 3D viewing ("Who Needs It?", 2008). Other challenges include the availability of 3D cameras and entertainment-industry technicians with stereoscopic production skills, and the overall costs of 3D filmmaking, which are about three times higher than recording and broadcasting in HD (Wood, 2010).

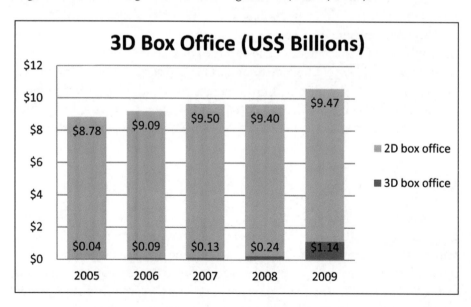

Source: Motion Picture Association of America, "Theatrical Market Statistics 2009," retrieved from http://www.mpaa.org/Resources/091af5d6-faf7–4f58–9a8e-405466c1c5e5.pdf

With improvement in the technology, however, sales of 3D TVs are expected to rise dramatically. Displaybank, a consumer-display market research group, reported in mid-2010 that 6.2 million sets would be sold globally that year, or 3% of all TVs, and predicted a market share of 31%, or 83 million sets, by 2014 (Displaybank, 2010).

Streaming Internet videos online is another venue where 3D technology is being actively researched by high-tech companies. The idea is to offer 3D movies to be played on devices that are already equipped with 3D viewing technology, such as video game consoles. Many companies, such as Samsung, already offer a TV app to view trailers for 3D movies (Katzmaier, 2010).

3D visualization software is also employed by professionals who work on product development and design. Architects, landscape and industrial designers, engineers, and medical professionals use the technology to visualize their concepts in 3D images (Leavitt Communications, 2001). 3D is also being explored as a marketing tool. Some companies have created virtual shopping applications that personalize an individual's body dimensions using 3D technology, which customers use to "virtually" try on clothes. The Ford Motor Company has its own virtual-world technology that allows buyers to design their own cars. Future 3D applications currently being explored include hand-held devices such as cameras, laptops, and camcorders.

Another emerging technology, holography is a technique that allows the light scattered from an object to be recorded and later reconstructed so that when an imaging system (a camera or an eye) is placed in the reconstructed beam, an image of the object will be seen even when the object is no longer present ("Holography," n.d.). As mentioned in Chapter 6, holography is being employed in telepresence systems for high-impact videoconferencing and virtual presence experiences.

Virtual Reality

Virtual reality (VR) technology, which produces an interactive 3D environment that simulates real-life or fictional settings, has existed since the 1990s. Participants in virtual worlds use avatars to inhabit and interact in the virtual space. High-speed Internet connections, powerful computer graphics cards, and audio designed for stereo headphones have combined to produce virtual worlds of sufficient depth and detail for deep user immersion.

Video games are one medium where younger players are discovering immersive VR environments. Only 2% of all computer gamers have visited a virtual world, but this percentage rises to 11% among teenage gamers. Six percent of gamers overall use some type of avatar (Lenhart, Jones, & Macgill, 2008). Teens are twice as likely to visit a site hosting a virtual world than people over 18 (Zickuhr, 2008).

Outside of gaming, virtual reality systems are being employed in the U.S. military to teach soldiers how to use particular weapons that might otherwise be too expensive to use for training. VR technology is also used to simulate combat situations for improving unit cohesion (LeCappelain, 2010) and combat decision-making (Huergo, 2005).

Social sciences researchers are using immersive VR to study health behaviors, such as sexual risk-taking, that would be too difficult and/or dangerous to reproduce with live subjects. It also gives researchers the ability to monitor actual responses rather than "intentional behaviors" (Krane, 2011). Medical applications include using VR as a stimulus for the repetitive-motion tasks necessary to regain mobility after a stroke or other brain injuries ("Stroke Recovery Goes 3-D," 2010).

In the workplace, virtual reality can be used for employee training. Virtual worlds like Second Life, for instance, can provide a forum that allows workers to meet, network, and collaborate with colleagues in other locations (Kopp & Burkle, 2010). Virtual reality can be particularly useful in training workers for tasks where the training itself could be dangerous. They can gain these basic skills safely in the virtual world before being exposed to the dangers of real-life work situations. Areas where this approach is being used include the military, law enforcement, firefighting, emergency medical response, dangerous driving, mining, railway operations, space exploration, aviation, marine exploration, hazardous materials handling, and working with nuclear energy.

Other less hazardous areas where VR is being used for training purposes include surgery, dentistry, engineering, forensics and accident investigation, welding, lathe operation, construction, stress-testing equipment, aircraft maintenance, biotechnology, and vehicle and aviation design (Ausburn & Ausburn, 2008). Workers can be trained to test and operate complicated machinery in a virtual setting, and they can repeat the lessons as many times as required, without any need to access the actual machinery and no risk to the learners or others around them (Blümel, Termath, & Haase, 2009).

As Web 2.0 technologies become an indispensable part of the lives of Generations X and Y, higher education institutions are looking to incorporate VR technology for learning and instruction. There are many prevailing theories and approaches to using VR in education. Some researchers believe pedagogical success can be achieved using VR because of its potential to form a community where everyone—instructor, peer, expert, and novice—can learn from one another (Bronack et al., 2008). Others cite ways to incorporate different multi-user, Web-based virtual learning environments to encourage role-playing and team-building activities (Livingstone & Kemp, 2008). Some researchers attribute significant pedagogical relevance to virtual worlds because they believe students will be more likely to explore, participate, discover new knowledge, and develop industry-relevant skills with greater intrinsic motivation and autonomy in these environments (Dreher, Reiners, Dreher, & Dreher, 2009).

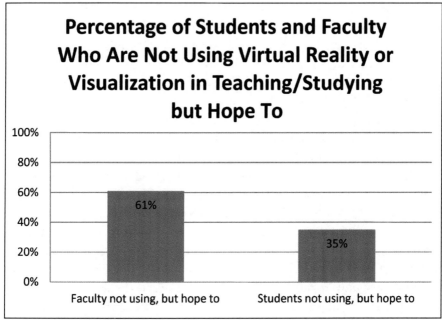

Source: "Technology, Teaching, & Learning Survey: Chapter 4. Technology Support for Learning-Centered Mission," by Lehigh University, retrieved from http://www.lehigh.edu/~infdli/teachtech/downloads/Chapter4.pdf

While VR research is gaining attention, actual adoption of virtual environments in higher education has been slower than anticipated (Collins, 2008). VR systems require sophisticated, integrated equipment and multiplatform capabilities, and many learner-focused institutions find the costs prohibitive. Another potential barrier is that VR technologies lack interoperability among different virtual-world platforms, making it difficult to plan and operate systems on a large scale. Last, issues of trust and other proprietary factors for their use and operation have to be successfully resolved before VR technology becomes mainstream. In spite of these obstacles, as both traditional and nontraditional students continue to show interest in virtual environments, education institutions will be more interested in exploring options that will help students feel more connected to others taking the course, while immersing themselves in a richer and more appealing learning environment (Collins, 2008).

Augmented Reality

Augmented reality (AR) defines the enhancement of an electronic image of the real world with computer-generated imagery and sounds for a hybrid sensory experience. A subset application of virtual reality research, AR systems are still in the early stage of development. But advancements in computer vision, object recognition, and other technologies are contributing to increasingly sophisticated and powerful AR environments, many of which require nothing more than a current-generation smartphone (Kroeker, 2010).

The main objective of AR systems is to make the virtual data indistinguishable from reality, which requires real-time free movement. Two main system-evaluation factors are (1) the update rate for augmenting the image, and (2) the accuracy of the registration (Vallino, 2006). Put together, these two elements provide "a live direct or indirect view of a physical, real-world environment whose elements are augmented by computer-generated sensory input, like sound or graphics" (Kime, 2011).

Equipment required for augmented reality practice includes a camera, a viewing device (webcam or camera-enabled phone), and a tracker or marker for projection. The chosen environment is projected on the screen, and digital data are overlaid upon the environment. Many kinds of headgear can also be used for projection (Mastrion, 2010).

Gesture-based computing is another example of AR in which humans interact with mechanical devices using gestures. Nintendo's Wii is considered an interactive rendering of augmented reality where the users' actions can be seen in both real and virtual worlds ("Augmented Reality in Education," n.d.)

According to figures from ABI Research, the market for augmented reality (AR) in the US alone is expected to hit $350m in 2014, up from about $6m in 2008, or, around 50 times more from 2008 to 2014.

(http://www.luxist.com/tag/Pew%20Research%20Center%20Millennials/)

Among current real-world applications, museums are using AR in innovative ways. The J. Paul Getty Museum, for instance, uses AR to allow visitors to view and learn about historical objects without touching them. London's Natural History Museum provides visitors with handheld devices that incorporate video technology so that visitors can view the dinosaurs around the actual space of the museum (Johnson, Smith, Willis, Levine, & Haywood, 2011).

Digital marketing companies have found creative uses for AR technology through virtual experiencing of products and gadgets. For instances, consumers are able to try on jewelry, watches, and clothes, allowing interactivity in the selling and buying of products (Kime, 2011).

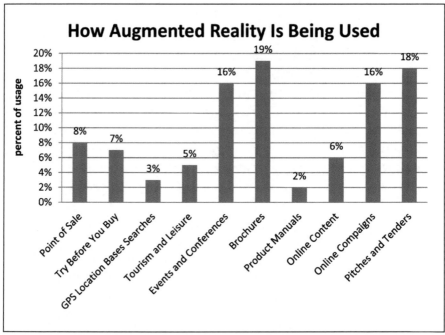

Source: "Marketing with Augmented Reality—An Infographic," by Hidden Creative, retrieved from http://www.slideshare.net/HiddenCreative/marketing-with-augmented-reality-an-infographic

Mobile augmented reality is quickly trending as the next revolution in AR technology, based on the different types of apps being developed for iPhones and Android-based smartphones. These embrace a variety of uses, from automotive repair and travel/tourism to real estate, advertising, and marketing (Sung, 2011).

Mobile AR apps now exist that use a combination of a camera, GPS, and social networking sites to provide instant information about people's Facebook status (The Astonishing Tribe) or that tag individuals via photos at a particular location in the world (Panoramio). The underlying technology here is identifying a position through a GPS system and using the Internet for additional information. App developers are already envisioning the next step, where face recognition technology and computer vision replace GPS technology for instant identification purposes (Sung, 2011).

These capabilities, of course, bring into question the idea of trust and privacy as apps become increasingly sophisticated and ubiquitous, allowing even strangers to instantly access information via social networking sites as they see each other face to face. AR specialists also reflect upon the downside of blurring of the virtual and real worlds, where having a cup of coffee with someone becomes a thing of the past, and technology subsumes the personal nature of interactions.

While the social element of AR is being debated, many experts see it as the future of education. In a recent report offered by technology and higher education consultants, AR was 1 of 10 technologies, along with mobile devices, e-readers, and gaming-based learning, predicted to penetrate the higher education realm within the next 2 years in impactful ways (Johnson et al., 2011). AR offers visual and highly interactive forms of learning, in which the layering of data over the real world in dynamic forms creates an active rather than a passive environment for students to take charge of their learning (Johnson et al., 2011).

Gaming

Video games, whether played on computers, home game consoles, portable gaming devices, or smartphones, continue to command the attention of players, the electronics and software industries, and advertising—both in the media and within the games themselves.

Video games can be played individually, by multiple users on the same device, online with a regular group of remote players, or with thousands of strangers across the Internet and around the world. Some online games are played in "real time" and represent more of a virtual community or world created by the game and its players, who often assume in-game avatars.

Buying or renting video games ranked seventh in a Nielsen survey of how U.S. households spent their entertainment budget. It ranked below going out to eat, subscribing to cable TV, and going to the movies, but above buying books and magazines, purchasing music, participating in sports, and renting movies. This meant that 4.9% of the average U.S. entertainment budget went to buying or renting video games and consoles. This figure rose to 9.3% in houses where video games were actively purchased (NielsenWire, 2010).

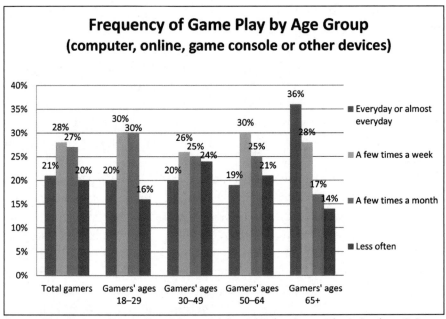

Source: "Pew Internet Project Data Memo: Adults and Video Games," by A. Lenhart, S. Jones, and A. R. Macgill, 2008, retrieved from http://www.pewinternet.org/~/media//Files/Reports/2008/PIP_Adult_gaming_memo.pdf.pdf

The computer gaming industry generated $10.5 billion of sales in 2009, and has grown almost 10% from 2005 to 2009 (Industry Facts, 2011). Technological advances such as more powerful processors, smaller portable consoles, and controllers that allow gamers to participate in physical activities, have sparked innovations in gaming software. The future of gaming lies in new advances that will allow for facial recognition and more integrated and voice-activated motions, digital distribution of new games, and augmented reality that melds the virtual world of the games with gamers' real-world locations (Snider, 2010).

A variety of multiplayer gaming options have evolved as technology has supported increasingly complex graphics and larger user bases. There are role-playing games; real-time strategy games; simulations, also called *sims*; rhythm games; management games; and social games, where the main objectives are to meet and socialize in avatar form. In management games the player has to perform some activity and achieve some goals. There is a wide variety of possible goals, such as running a hotel or restaurant, babysitting, working at a gas station, being a real estate broker, running an animal shelter, managing an ocean park, running a dress shop, and so on. Hundreds of these games are available for free online.

Businesses and employers are beginning to recognize the potential of using games as resources in the workplace. For example, one company that wanted to bring managers together in support of corporate changes used a simulation game during a managers' conference. The game portrayed a virtual company that was facing the same challenges as the actual company. Participating managers acted as consultants to meet the challenges of their fictitious companies. In doing so, they came to understand the necessity for change and the strategies the company's top management was using (Photo_iStockphoto.com, 2010).

Games are also being used to improve work performance in certain jobs. A study at Beth Israel Medical Center in New York found that surgeons who use video games to warm up (which improved their fine motor coordination) performed laparoscopic surgery faster and more accurately than those who did not. The military uses "war games to train soldiers for combat and simulators to assist foot soldiers in differentiating between combatants and noncombatants in urban settings" (Perkins, 2009). In 2004 the U.S. Army announced that it was developing a massively multiplayer simulation game called AWE (asymmetric warfare environment) to train its personnel for urban warfare ("Massively Multiplayer Online Game," n.d.). Commercial and military pilots both receive some of their training in flight simulators (Perkins, 2009). Using flight simulators is obviously safer for the new pilots and for people on the ground, and more cost effective than risking extremely expensive planes.

In the business context, video games have been shown to be a successful platform for advertisements via product placement. A study conducted by The Nielsen Company, along with game publisher Electronic Arts and Gatorade, found that gamers spent 24% more on Gatorade products after seeing the brand and products in games that they played (Guzman, 2010).

Another creative use of gaming in the business world has been to use them as a form of marketing, through which the consumer is having fun and not aware of receiving a sales pitch. For example, Disney released a game called RhinoBall as a free game for the Apple iPhone and iPod Touch to advertise its movie *Bolt*. The game involved manipulating characters from the animated movie (toucharcade.com). Likewise, Honda promoted an online racing game that introduced players to its cars (Perkins, 2009).

Popular Education Games

- *Civilization* and *Rise of Nations* are intended to teach about how civilizations and nations evolve and grow. Students learn about social, political and historical models by playing the games. For older students, these games can teach skills like communication, spatial literacy, and teamwork.

- *EVE Online* requires players to learn about economics, physics, teamwork, long-term planning, and communication.

- *Second Life* encourages players to write about their shared experiences in blogs and posts. In virtual worlds like *Second Life*, players can actually learn to construct the worlds themselves, picking up valuable technical skills in the process. (Alexander, 2008)

Higher education institutions are also investigating how they can engage students through gaming technology. Constance Steinkuehler, an assistant professor of educational communication and technology at the University of Wisconsin, argues that students learn very important skills when playing games like *World of Warcraft*. They become deeply engaged in the play and put a lot of work into supporting it. For example, Steinkuehler explains that World of Warcraft players were gathering data on a monster in the game, charting that information on Excel, and building mathematical models that could help them beat the monster. In this process they were clearly learning important and useful skills. She points out that this game teaches players how to solve complex problems and involves them in collaborative learning (Young, 2010).

With the growth of virtual campuses and increased enrollment in online learning programs, a big potential exists for using gaming technology in student education. One group of experts proposes that educational institutions add more virtual space for peer-learning and socializing. They suggest these interactive spaces be based on the model of massively multiplayer online games (Preston, Booth, & Chastine, 2004).

Games as a teaching tool have not just become a part of the curriculum; they are now the subject of academic inquiry. There are professional conferences for experts in the field, peer-reviewed articles, books, educational programs, and faculty positions. As scholars and researchers receive grants to study games, test the effectiveness of using them in education, and develop new approaches to such educational gaming, we can expect to see rapid changes in the relationship between gaming and education.

Games clearly can teach a variety of information and skills, such as dealing with money, creating a business strategy, scheduling and planning ahead, being persistent, and many other life skills—all while players have fun and hardly notice that they are learning. When it involves other players, the immersive experience also offers a whole new way to socialize. But how will these experiences of gaming affect general lifestyles?

Glen Heimstra, a self-denominated futurist and visiting scholar at the University of Washington Human Interface Technology Lab, thinks it will have a major impact. He sees the distinction between gaming and life becoming smaller and smaller, until it becomes possible to get real and virtual life confused. In 25 years he anticipates the use of virtual retina display glasses or contact lenses that will create augmented reality screens. In other words, reality will be seen through the screen of virtual reality. An Internet connection will be something the person can wear at all times (Jones, 2007).

Heimstra predicts that digital natives, those in Generations Y and Z who have grown up with this kind of gaming, will demand a different kind of world. Heimstra puts it interestingly: "You've heard the term the 'digital natives?' It's a great term. Baby boomers sort of invented microchips. Gen X invented the World Wide Web. And the digital natives, they live in that world (Jones, 2007, p. 4)."

Digital natives learn technology like games in a few minutes and are used to doing several things at once. A virtual interface with the real world would seem normal to them. Heimstra suggests that digital natives will take information technology to places older generations have not even dreamed of.

With gaming evolving at a fast pace and more digital natives entering the workforce every day, we can expect to see more use of gaming as a training and social technology for workers and as an expanded form of advertising, particularly for products aimed at Generation Y. We can also expect to see many new applications for gaming that have not even been thought up yet.

Robotics

Robots have been a big part of science fiction for a long time. In the real world, robotics is the field of science that develops and modifies machines to perform an assortment of tasks with varying degrees of autonomy. This technology is being envisioned, and in many cases today, deployed in myriad ways. For instance, one day robots may not only take people's places at company meetings in specially designed rooms, but "walk" around to view or inspect remote locations, indoors and out. Robotics technology is currently being employed in such life-saving circumstances as search-and-rescue missions, trauma centers, and military battlefields.

Robots began to capture peoples' imagination in the early part of the 20th century when the first humanoid robot was exhibited by the Westinghouse Electric Corporation at the 1939–1940 World's Fair. The first autonomous robot, Elsie, was created by William Grey Walter at Bristol University in 1948. The first industrial robot was patented in the 1950s by George Devol. Today, robots and robotics technology are being used successfully in manufacturing and production, particularly for tasks that require precision, accuracy, and repetition (Hall, 2010).

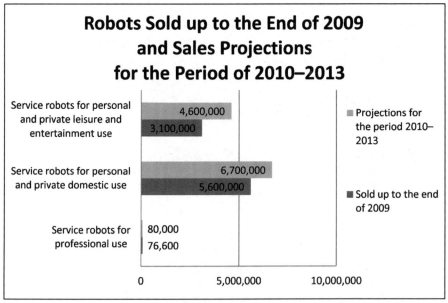

Source: "World Robotics 2010 Service Robots: Service Robotics Statistics," International Federation of Robots, retrieved from http://www.ifr.org/service-robots/statistics/

The robotics industry is divided into two sectors: service robots and industrial robots. The year 2008 was a banner year for industrial robot sales, but there was a 47% decline (60,000 units) in 2009 because of the global economic crisis. In 2010, the market recovered by 27%, or around 76,000 units (International Federation of Robots [IFR], 2011). Japan and Korea are the main consumers of industrial robots in Asia. Other major robotics markets are the United States, Germany, and Italy. Consumer markets are expanding in China, India, Brazil, and Russia, which partially accounts for the recovery in 2010 (IFR, 2011).

The IFR estimates that there are over 500,000 operational industrial robots in Asia, 330,000 in Europe, and 180,000 in North and South America (IFR, 2010). Automakers are the largest consumers of industrial robots; they purchased 36% of those sold in 2009, and the health of the robotics industry partly depends on automakers' vigor due to their share of the market. Electronics industries accounted for 18% of industrial robot purchases. The rubber and plastics industries bought 10% of them that year, and the metal and machinery industries purchased 9%. The food and beverage industry purchased 5% of the industrial robots sold in 2009 (IFR, 2010).

In 2009, the total value of professional service robots was $13.2 billion. Thirty percent of service robots are used for military applications; other uses include medicine, agriculture (milking), cleaning, construction and demolition, search and rescue, and security logistics.

Personal robots, used as toys or for teaching or cleaning, are still a very small portion of the market. The demand for robots that can assist handicapped people with personal care and mobility is small but growing (IFR, 2010).

Although robots can drastically reduce human beings' need to perform repetitive physical tasks, they also have societal significance for education and work. Robots may threaten to displace or eliminate jobs in industries dominated by physical labor, but they also offer educational and workforce opportunities. The field of robotics may generate a host of new jobs related to development, design, and maintenance of these devices, as well as in educating robot operators on their efficient use.

Robots are also being used for newly developed purposes. These include food processing, mail handling, nanotechnology, military purposes, and surgery. The first surgical robots were introduced in 1999, and robot-assisted surgery has become standard procedure in some areas of medicine. Robots allow surgery to be less invasive, which makes it easier for the surgeon (once he or she learns the technique) and easier on the patient. In 2008 about half of all prostate cancer surgery in the U.S. was done with the help of robots. About 400 institutions outside the U.S. also have the robot technology for this procedure (Von Drehle, 2010). Though robotic surgery is currently expensive, it can produce shorter hospital stays, fewer infections, and shorter recovery time, all of which can save on healthcare costs. It will take a careful analysis, however, to determine if robotic surgery is cost effective (Hall, 2010).

There has long been a concern that robots would put people out of work; after all the job losses in the recent recession, the idea of eliminating jobs is a very sensitive one. At least one commentator, Owen Hall, claims that the loss of manufacturing jobs in developed countries with higher wages is inevitable in the current marketplace. Over 2 million of these jobs were lost from 2001 to 2007, and an additional 41,000 disappeared in 2009 alone. Many of these jobs were lost because of offshoring, where U.S. companies opened manufacturing facilities in areas with lower labor costs. U.S. companies argue that they have to move offshore in order to stay competitive in the global marketplace with overseas companies that have low labor costs. Hall argues that robotics, far from causing Americans to lose jobs, may actually secure some of them. If U.S. firms can be competitive by using robotic manufacturing in the U.S., then these companies, with all their support needs, will remain in the country. Some types of jobs will be lost—they will be lost anyway—but many other jobs will be saved. Hall points out that Japan has already begun to use this approach with some success. He sees education as crucial to using this technology effectively and sustaining America as a player in the global market of the future (Hall, 2010).

Robots are also taking on a larger role in workplace telepresence technology. These so-called telepresence " bots" can attend meetings that are held in special rooms. They are operated by meeting participants in other locations, and the bots are those people's eyes, ears, and mouths. Other participants in the meeting talk to the remote participants by talking to the bot. The current bots used for this purpose weigh about 35 pounds and have a neck that can extend from three feet to about six feet. One person described them as a Skype-cam on a stick. These bots can be used in any environment where the robot can maneuver. Companies have reported mixed results so far when using this technology ("Bots in the Boardroom," 2010). Talking to a little robot with someone's face showing on the screen may take some getting used to. It doesn't have the familiarity of meeting in a telepresence room, but at this point it is more cost effective and versatile.

Summary

Immersion, gaming, and robotics have not only captured people's imagination for augmenting real-life experiences, but have found their place in society for their ability to help users accomplish dangerous tasks, enhance learning environments, and permit training and learning for everyday jobs and skills. The reduction of costs, and the rising efficiency of technology platform integration, will determine the scope for the wider use and future growth of these technologies.

References

Augmented reality in education. (n.d.). In *WikEd*. Retrieved June 23, 2011, from http://wik.ed.uiuc.edu/index.php/Augmented_Reality_in_Education

Ausburn, F. B., & Ausburn, L. J. (2008, October). Send students anywhere without leaving the classroom: Virtual reality in CTE. *Techniques*, 43–46.

Bell, M. W. (2008). Toward a definition of "virtual worlds." *Journal of Virtual Worlds Research*, 1(1). Retrieved from http://journals.tdl.org/jvwr/article/view/283/237.

Blümel, E., Termath, W., & Haase, T. (2009, May). Virtual reality platforms for education and training in industry. *iJAC*, 2(2), 4–12.

Bots in the boardroom. (2011, July 1). *Credit Union Magazine*. Retrieved from http://www.creditunionmagazine.com/articles/bots-in-the-boardroom.

Bronack, S., Sanders, R., Cheney, A., Riedl, R., Tashner, J., & Matzen, N. (2008). Presence pedagogy: Teaching and learning in a 3D virtual immersive world. *International Journal of Teaching and Learning in Higher Education*, 20(1), 59–69. Retrieved from http://www.isetl.org/ijtlhe/pdf/IJTLHE453.pdf

The champions of 3D. A look behind the big names behind the 3D revolution (2011, June 6). Retrieved from http://www.virtualgeneration.net/en/features/3D+Cinema+and+TV/9978/1/The_Champions_Of_3D/

Collins, C. ("FLEEP TUQUE"). (2008, September/October). Looking to the future: Higher education in the metaverse. *EDUCAUSE Review*, 43(5). Retrieved from http://www.uh.cu/static/documents/RDA/Looking%20to%20the%20future.pdf

Displaybank. (2010, May 17). 6.2 million 3D TVs expected to be sold globally in 2010 showing CAGR of 91% until 2014. Retrieved from http://www.displaybank.com/eng/info/sread.php?id=5750

Dreher, C., Reiners, T., Dreher, N., & Dreher, H. (2009). Virtual worlds as a context suited for information systems education: Discussion of pedagogical experience and curriculum design with reference to Second Life. *Journal of Information Systems Education*, 20(2), 211–224.

Eskelsen, G., Marcus, A., & Ferree, W. K. (2009). *The digital economy fact book* (10th ed.). Washington, DC: The Progress & Freedom Foundation.

Guzman, G. (2010, September 14). Video game advertising: Playing to win . . . and sell. Retrieved from http://blog.nielsen.com/nielsenwire/consumer/video-game-advertising-playing-to-win%E2%80%A6-and-sell

Hall, O. P. (2010). Is robotics America's ticket to continued global competitiveness? *Graziadio Business Report: A Journal of Relevant Information and Analysis, 13*(1).

Holography. (n.d.). In *Wikipedia*. Retrieved June 23, 2011 from http://en.wikipedia .org/wiki/Holography

Huergo, J. (2005). Virtual reality, real ingenuity [Press release]. Office of Naval Research website: http://www.onr.navy.mil/Media-Center/Press-Releases/2005/Virtual-Reality-Real-Ingenuity.aspx

Immersion (virtual reality). In *Wikipedia*. Retrieved June 23, 2011 from http://en.wikipedia .org/wiki/Immersion_%28virtual_reality%29

Industry Facts. (2011). Economic data. Retrieved from Entertainment Software Association website: http://www.theesa.com/facts/econdata.asp

International Federation of Robotics. (2010, January 27). *Executive summary of world robotics 2010: Industrial robots and world robotics 2010: service robots*. Retrieved from http://www. worldrobotics.org/downloads/2010_Executive_Summary_rev%281%29.pdf.

Johnson, L., Smith, R., Willis, H., Levine, A., & Haywood, K., (2011). *The Horizon report: 2011 edition* Austin, TX: The New Media Consortium. Retrieved from New Media Consortium website: http://www.nmc.org/pdf/2011-Horizon-Report.pdf

Jones, S. (2007). The gaming world of 2025: Futurist, Glen Hiemstra, predicts a world inhabited by digital natives. *Bloomberg Businessweek*. Retrieved from http://www.businessweek .com/innovate/content/feb2007/id20070213_645807.htm

Katzmaier, D. (2010, March 12). 3D TV FAQ. Retrieved from http://news.cnet.com/3d-tv-faq/#1

Kime, S. (2011, February 1). Engaging the millennial: Augmented reality and the wired generation. Retrieved from http://www.luxist.com/2011/02/01/engaging-the-millennial-augmented-reality-and-the-wired-generat/

Kopp, G., & Burkle, M. (2010, August). Using Second Life for just-in-time training: Building teaching frameworks in virtual world. *iJAC, 3*(3), 19–25.

Krane, B. (2011, February 24). Conducting virtual reality research. *U Conn Today*. Retrieved from http://today.uconn.edu/?p=10278.

Kroeker, K. L. (2010). Mainstreaming augmented reality. *Communications of the ACM, 53*(7), 19–21.

Leavitt Communications. (2001, November). 3D technology: Ready for the PC? Retrieved from http://www.leavcom.com/ieee_nov01.htm

LeCappelain, J. (2010, September 20). FITE demonstration builds small unit teamwork, cohesion. Retrieved from United States Joint Forces Command website: http://www.jfcom. mil/newslink/storyarchive/2010/pa092010.html

Lenhart, A., Jones, S., & Macgill, A. (2008, December 7). *Pew Internet Project data memo: Adults and video games*. Retrieved from Pew Internet & American Life Project website: http://www.pewinternet.org/~/media//Files/Reports/2008/PIP_Adult_gaming_memo.pdf.pdf

Livingstone, D., & Kemp, J. (2008). Integrating web-based and 3D learning environments: Second Life meets Moodle. *UPGRADE, 9*(3), 8–14. Retrieved from http://www.cepis.org/ upgrade/files/2008-III-kemp.pdf

Massively multiplayer online game. (n.d.). In *Wikipedia*. Retrieved June 23, 2011 from http://en.wikipedia.org/wiki/Massively_multiplayer_online_game

Mastrion, G. (2010). Augmented reality: The new, new media. *Pharmaceutical Executive*, *30*(7), 82–83.

NielsenWire. (2010, February 22). Video games score 5% of U.S. household entertainment budget. Retrieved from http://blog.nielsen.com/nielsenwire/consumer/video-games-score-5-of-u-s-household-entertainment-budget

Perkins, B. (2009, November). World of Warcraft in the workplace. *Computerworld*, *43*(32), 30.

Photo_iStockphoto.com. (2010, August). Workplace Rx: Game's on. Retrieved from http://www.istockphoto.com

Preston, J. A., Booth, L., & Chastine, J. (2004). Improving learning and creating community in online courses via MMOG technology. Retrieved from http://www.scribd.com/doc/210290/MMOG-in-Onlinel-Education

Snider, M. (2010, January 5). A look at the future of video games. Retrieved from http://content.usatoday.com/communities/gamehunters/post/2010/01/a-look-at-the-future-of-video-games/1

Sniderman, Z. (2011, February 7). How does 3D work? Retrieved from http://mashable.com/2011/02/07/how-does-3d-work/

Statistics. (2011, January 27). Retrieved from International Federation of Robotics website: http://www.ifr.org/industrial-robots/statistics

Stroke recovery goes 3-D: Canadian video game takes rehab to the next level. (2010, June 21). *ScienceDaily*. Retrieved from http://www.sciencedaily.com/releases/2010/06/100607165617.htm

Sung, D. (2011, March 4). Augmented reality in action: Social networking. Retrieved from http://www.pocket-lint.com/news/38918/augmented-reality-social-networking-dating

Vallino, J. (2006, September 3). Augmented reality page. Retrieved from http://www.se.rit.edu/~jrv/research/ar/

Von Drehle, D. (2010). Meet Dr. Robot. *Time*, *176*(24), 44–49.

Who needs it? Three-dimensional television is coming, whether you want it or not. (2010, April 29) *The Economist*. Retrieved from http://www.economist.com/node/15980777?story_id=15980777

Wood, D. (2010, November). TV: HD to 3D. *Televisual*, 31–40.

Young, J. R. (2010, January 4). 5 teaching tips for professors—from video games. *The Chronicle of Higher Education*. Retrieved from http://chronicle.com/article/5-Lessons-Professors-Can-Learn/63708/

Zickuhr, K. (2010, December 16). Generations 2010: Online activities. Activities that are most popular with teens and/or millennials. Retrieved from http://pewinternet.org/Reports/2010/Generations-2010/Activities/Younger-cohorts.aspx

SECTION IV

HIGHER EDUCATION
AND IMPLICATIONS

This section ties together our focus on work, society, and technology in a discussion within the context of higher education requirements. We define the importance of learning environments and how personal learning ecosystems define many of the information-gathering processes that take place today. We present the job skill characteristics that will be vital in the future, as well as their implications for important stakeholders in individuals, businesses, and the government. We also offer higher education models, and ask several critical questions:

- Who are the students of today and tomorrow?

- What are their anticipated educational expectations?

- What new educational models and approaches will be required to meet future societal needs?

- What is the role of higher education institutions, business, and government in the context of education and work?

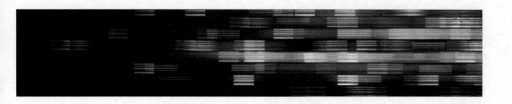

· 8 ·

LEARNING ENVIRONMENTS

Individualized Methods for Gathering Information

Technology is constantly in motion, and by the time this book reaches the market many of the technologies discussed in this book will be outdated. In *.edu* (Wilen-Daugenti, 2009), several technological trends—namely mobile, Web 2.0, and social technologies, video and gaming—were reviewed and discussed in the context of their impact on higher education and how they were being used to create personalized learning environments. Today, these technologies have advanced further, and learners are beginning to embrace additional modalities of learning, namely, *mobility, collaboration*, and *immersion*, which have opened up a sea of possibilities and expectations.

At the same time, the proliferation of innovative technologies can often overwhelm individuals in their quest for information. Information overload and multitasking frequently lead to confusion and heightened stress, both in society and in higher education. However, it is important to note that while modern technology promotes the use of a complex array of learning options, individuals are also uniquely positioned to select what works best for them and to create personal learning ecosystems based on their preferred method of gathering, processing, and acting upon information.

This chapter will focus on learning environments and their evolution in real-life circumstances. By use of a personal example, I will illustrate how individuals construct personal learning ecosystems based on their level of expertise and comfort level, and in turn, how these learning modes transfer to society, work, and higher education.

Learning Environments

Learning environments result from the blurring of lines between traditional, face-to-face learning in classrooms and the ever-expansive nature of technology-enriched resources, such as the mobile, collaborative, and immersive experiences that we have at our disposal today.

In learning environments, learners choose how they access information, which resources they use, and where they obtain the knowledge they need to be academically successful. Individuals in such environments have unrestricted access to all information resources, both in the physical and virtual world; in effect, physical barriers to learning and research no longer exist. A flexible learning environment allows learners to customize their experience to their unique needs and preferences.

One example of a learning environment is the use of the Internet. People use the Internet for daily life decisions. In fact, they have become increasingly dependent on the Internet, as evidenced by Pew researchers (Boase, Horrigan, Wellman, & Rainie, 2006), who noted that the Internet was used by:

- 54% of adults in helping someone cope with an illness;
- 50% of adults in career training;
- 45% of adults making financial decisions;
- 43% of adults looking for a home;
- 42% of adults in deciding on a school or college for themselves or children;
- 24% of adults in buying a car; and
- 14% of adults for switching jobs.

In these instances, individuals effectively use the Internet to create a personal ecosystem of knowledge acquisition that otherwise would have to be absorbed through more traditional and physical means, such as a classroom, book, newspapers, brochures, and so forth.

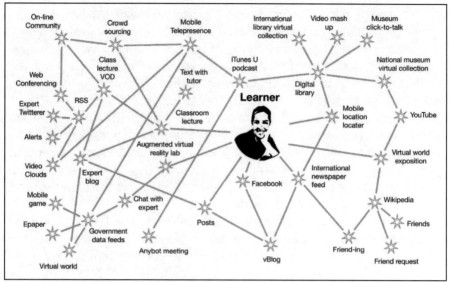

Fig. 8.1 Individualized Learning Environments

Try recollecting a recent major life event, such as a wedding, college selection, buying a house or car, or addressing a health issue, where credible information needed to be collected in a compressed period of time. Then think through how you went about gathering information and learning about the subject. Then observe how others might also approach informing themselves about the same subject.

Singular Approaches to a Common Goal: A Personal Encounter with Information Choice

This example demonstrates multiple approaches that can be adopted for a single quest based on individual knowledge preferences, technical aptitude, or personal choice.

My mother-in-law was diagnosed with breast cancer in 2007. It was unexpected and took her and the family by surprise. She had limited knowledge and access to Internet, electronic materials, and lived too far from her local library. She was somewhat home bound and has limited use of technology—a phone. Her learning environment or access to information was limited to a medical manual that was dated, her doctor, local family, and whomever she could reach via phone or talk to at church. She reached out to family for answers. We in turn felt obligated to quickly educate ourselves on the topic of breast cancer so that we could support her and talk with her to her physicians.

My personal learning environment entailed investigating her physicians' credentials, including the radiologist, the surgeon, and her family practitioner. I did this by searching online medical databases, university databases, and official hospital websites, which are freely available on the Web. I also went to the top university medical websites to learn more about the cancer, stages, and statistics on survival as well as treatment options. I also visited communities where cancer survivors submitted their experiences and advice as well as recommendations. I read recent alerts and updates on various medical blogs so I could keep current on topical issues. I subscribed to numerous journals and RSS feeds and used my laptop and mobile device, not to mention reading the materials and electronic books on my e-book reader.

My husband took a different path. He was not as well versed on what resources were available on the topic so he typed the query "What is breast cancer?" into the search engine Google. A variety of resources populated indicating pages varying from a definition in Wikipedia and Webopedia to medical sites such as WebMD.com, Medline.com, as well as physician sites. He reviewed the various sites and eventually found himself overwhelmed with the number of resources, not knowing quite which ones to trust and which ones were accurate. He decided that he would pursue "ask the expert" sites such as www.justanswer.com and www.kasamba.com where you can review experts' bios and pay by the minute for online real-time advice that you choose to use.

Payment is based on customer satisfaction and experts are ranked by users. My husband found quite a few medical experts who helped guide him quickly through the question process and to educational resources that might help him understand the cancer, understand what his mother might be dealing with, and what type of questions to ask her physician. He also has a very robust social network on a number of sites and reached out online to a variety of friends who had mothers, aunts, or sisters who had experienced the same disease.

Within 24 hours the two of us, while keeping in contact online, texting, and chat, pursued very different learning paths and had accumulated most of the knowledge we needed. In addition, we both tapped into a variety of online as well as offline resources to accomplish our goals. Furthermore, each individual took a unique approach to the learning process. Within 48 hours we shipped to his mother a package of reading materials, and we set up phone conference calls with her, us, and her physicians to help her understand the disease and what would happen over the next few weeks. She felt relieved.

A 2010 Pew Research survey (Fox, 2011) of information-gathering practices on health matters showed the following. Of the 74% of adults who used the Internet:

- 80% of Internet users have looked online for information about any of 15 health topics such as a specific disease or treatment. This translates to 59% of all adults.

- 34% of Internet users, or 25% of adults, have read someone else's commentary or experience about health or medical issues on an online news group, website, or blog.

- 25% of Internet users, or 19% of adults, have watched an online video about health or medical issues.

- 24% of Internet users, or 18% of adults, have consulted online reviews of particular drugs or medical treatments.

- 18% of Internet users, or 13% of adults, have gone online to find others who might have health concerns similar to theirs.

- 16% of Internet users, or 12% of adults, have consulted online rankings or reviews of doctors or other providers.

- 15% of Internet users, or 11% of adults, have consulted online rankings or reviews of hospitals or other medical facilities.

Similarly, of those who use social network sites (62% of adult Internet users, or 46% of all adults):

- 23% of social network site users, or 11% of adults, have followed their friends' personal health experiences or updates on the site.

- 17% of social network site users, or 8% of adults, have used social networking sites to remember or memorialize other people who suffered from a certain health condition.

- 15% of social network site users, or 7% of adults, have obtained any kind of health information on the sites.

Personal learning environments can be highly diverse and layered, producing the necessary framework for a progressive mastery of a topic of interest. A student or an adult at work can choose how to gather information based on accessibility, availability, and familiarity of technological tools and services.

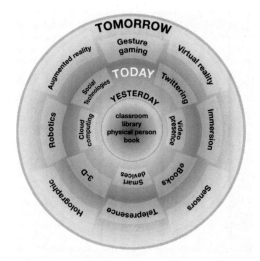

Fig. 8.2 Evolution of Information and Learning Resources

Figure 8.2 shows the evolution of information and learning resources. The circle in the center represents where many people have had their college learning experiences in the past. Here, traditionally, resources were limited to a classroom space, the physical presence of professors on a college campus, and the use of books and paper materials. In other words, learning was confined to what a professor could bring into the physical confines of the classroom or what the library had on its shelves.

The circles that span outside of the center show the progression of resources that have become available in recent years. These resources (video sharing, social networking sites, electronic books, etc.) complement what already exists and help learners expand their "pie" of available resources. Unlikein the past, not all of the resources readily available today are physical.

As we follow the circles to the outermost rim, we see additional resources that learners use, and how, given access, they can make the most of an abundance of resources to help customize their learning and information-seeking tasks.

Learners tend to select resources that best suit their personal preference. A key element to the learning environment model, therefore, is "choice" in where resources are utilized based on personal learning styles and preferences (paper, in person, video, virtual, etc.). Technology has greatly expanded these choices and empowered individuals in their quest for knowledge.

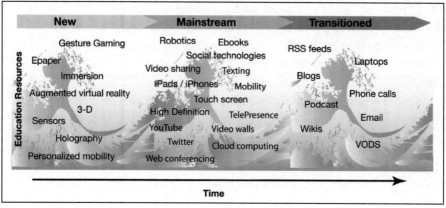

Fig. 8.3

Figure 8.3 depicts the transition of information resources, from left (newest) to right (oldest). On the right side are traditional information resources such as laptops, phone calls, email, video-on-demand (VODS), RSS feeds, blogs, podcasts, wikis, and so forth. These information resources still exist and continue to be in use today.

The center section indicates resources that have entered the environment in recent years and that are also being actively adopted today. In some cases the center resources have substituted the resources on the right, such as using Twitter instead of blogs, iPads and phones instead of laptops, video sharing over VODS, or virtual museums in place of virtual worlds. In other instances, new resources enhance existing resources by providing more visual, mobile, and collaborative options for learning.

On the far left are emerging resources such as gesture gaming, immersion, epaper, augmented virtual reality, 3D and holography, personalized mobility, and sensors. These are in the exploratory stage; some have been adopted, and others are used in conjunction with separate resources.

The flow of technology use, metaphorically speaking, thus comes and goes in waves, taking users to different forms and configurations but ultimately bringing them back to what feels most comfortable and offers the best outcome for the task at hand.

Summary

In a learning environment, be it education, work, or society, individuals choose how to learn, which resources to use, and where to obtain the knowledge necessary to be successful. Students in a learning environment have unrestricted access to all knowledge resources, in both the physical and virtual worlds, so barriers to learning and research no longer exist. In a flexible learning environment, moreover, learners can customize their learning experience to fit unique needs and preferences and enhance their overall perspective. It is often said that knowledge is power, and in today's knowledge gathering ecosystems, that power is being realized through uniquely personalized methods and resources tied to the individual's comfort level and their capacity to absorb and process information.

References

Boase, J., Horrigan, J., Wellman, B., & Rainie, L. (2006, January 25). The strength of Internet ties. Retrieved from http://www.pewinternet.org/~/media//Files/Reports/2006/PIP_Internet_ties.pdf.pdf

Fox, S. (2011). The social life of health information, 2011. Retrieved from http://www.pewinternet.org/~/media//Files/Reports/2011/PIP_Social_Life_of_Health_Info.pdf

Wilen-Daugenti, T. (2009). .edu: Technology and learning environments in higher education. New York: Peter Lang.

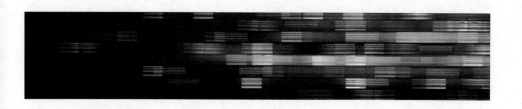

FUTURE OF HIGHER EDUCATION

While it is often difficult to predict with great certainty where higher education is headed, in May 2011 the Institute for the Future and the University of Phoenix Research Institute published *Future Work Skills 2020* in response to the growing demand for such insights by individuals, educators, business, and the government (University of Phoenix, 2011). It identified several drivers of change with the potential to reshape the future landscape that were considered most important and relevant to acquiring future work skills:

- *Extreme longevity:* Increased lifespans will change the nature of careers and learning. It is estimated that by 2025, the number of Americans over the age of 60 will increase by 70%.

- *Rise of smart machines and systems-workplace automation:* Human workers will be nudged out of rote, repetitive tasks. Smart machines will be integral to every domain of our lives, such as medicine, teaching, production, security, and combat.

- *Computational world:* The increased use of sensors and greater processing power make the world a programmable system. Every object that we come into contact with will be converted into data—and on an extreme scale. The ability to mine, manipulate, and interact with the data will become increasingly important.

- *New media ecology:* New communication tools require a new media literacy beyond text. As technologies such as video, digital animation, augmented reality, and gaming become more sophisticated and pervasive, the need for a new sensibility to use these technologies becomes more acute.

- *Superstructure organization:* Social technologies drive new forms of production and value creation. New technologies and social media platforms are driving an unprecedented reorganization of how we produce and create value.

- *Globally connected work:* Increased global interconnectivity puts diversity and adaptability at the center of organizational operations. There will be greater exchange and integration across geographic borders. (University of Phoenix, 2011)

In addition, the study also identified a range of key skills that would be relevant for the future:

- *Sense-making:* Determining the deeper meaning or significance of what is being expressed. In a world that is increasingly automated, there will be a need for new skills that can make sense of the automation and understand the output of a highly mechanized environment.

- *Social intelligence:* The ability to connect to others in a deep and direct way, to sense and stimulate reactions and desired interactions. This refers to the skill of detecting and absorbing people's emotions, gestures, and words in a highly complicated and automated environment. Demand for these skills is increasing as people in the world have become more distant. The ability to build collaboration, relationships, and trust is a key factor.

- *Novel and adaptive thinking:* Proficiency at thinking and coming up with solutions and responses beyond the rote or rule based. This refers to the skills of responding to unique and expected circumstances of the moment, and possessing the talent to adapt and provide novel thinking.

- *Cross-cultural competency:* The ability to operate in different cultural settings. This refers to operating in a globally connected world that requires teaming and collaboration at a distance—often across national borders.

- *Computational thinking*: The ability to translate vast amounts of data into abstract concepts and to understand data-based reasoning. As data increase, so does the need for people who can manipulate and understand it and be able to construct value and meaning from it.

- *New media literacy*: The ability to critically assess and develop content that uses new media forms and to leverage these media for persuasive communication. In a world that is increasingly full of rich media, this refers to the skill of deciphering and using multiple forms of such media (e.g., videos, mobile, and social technologies).

- *Transdisciplinary literacy*: The ability to understand concept across multiple disciplines. In an ever complex world the need for people who have deep and broad skills increases to be able to cope with multidisciplinary problems.

- *Design mindset*: The ability to represent and develop tasks and work processes for desired outcomes. This refers to developing the skill to design one's environment to be conducive to thinking rather than a traditional structured approach.

- *Cognitive load management*: The ability to discriminate and prioritize data by importance and to understand how to maximize cognitive functioning using a variety of tools and techniques. This refers to how people filter and focus on what is important in an ever-changing, complex, multimedia world.

- *Virtual collaboration*: The ability to work productively, drive engagement and determine presence as a member of a virtual team. (University of Phoenix, 2011)

Current Trends in Job Skills and Education

To assess the most vital future skills for the jobs of tomorrow, we must look more closely at where we stand today, and at where higher education enterprises are headed as they adapt to changing work and family structures. Today, employers are increasingly requiring degrees as a prerequisite to employment. Between 1973 and 2008, the share of U.S. jobs that required some training past high school climbed from 28% to 59%. By 2018 this number is projected to reach 63% (Carnevale, Smith, & Strohl, 2010). In earlier chapters, we discussed how trends such as globalization, changing demographics and family dynamics, and technological advances are transforming society in fundamental ways. Here, we discuss the trends that impact individuals' approaches to education and the factors that influence higher education institutions as they try to adapt to societal changes.

Trends Impacting Jobs of the Future

Economic shifts after the recession of 2007–2009 have remade the educational requirements in the U.S. workplace. Companies have moved jobs requiring only a high school education or less offshore to less educated workforces in developing countries, and many low-skill jobs that remain in the U.S. have been automated. These include jobs in farming, fishing, and forestry, as well as in manufacturing. Hundreds of thousands of these jobs have already been lost in the recession's wake, and it is predicted that 637,000 of those jobs will permanently disappear from the U.S. job market by 2018 (Carnevale et al., 2010). American companies are expected to develop more jobs, but most will require higher education, with an increasing number requiring a four-year degree. Unless workers continue their educations, an estimated 60 million Americans may be barred from reaching the middle class, relegated to a shrinking market of low-paying jobs or unemployment as the U.S. economy continues to recover (Carnevale et al., 2010).

At the same time, a shortfall of trained workers is anticipated. According to the Georgetown University Center on Education and the Workforce, the U.S. workforce will require 22 million new employees equipped with college degrees by 2018, but at the current rate it will fall short by at least 3 million. To meet this demand for an educated workforce, colleges and universities would have to increase the number of degrees awarded by 10% per year, a potentially monumental task (Carnevale et al., 2010).

Another study indicates that U.S. colleges and universities need to produce one million more graduates per year by the year 2020 to provide sufficient workers for the country to remain economically competitive. This will either require additional funding or more efficient provision of education (Auguste, Cota, Jayaram, & Laboissière, 2010).

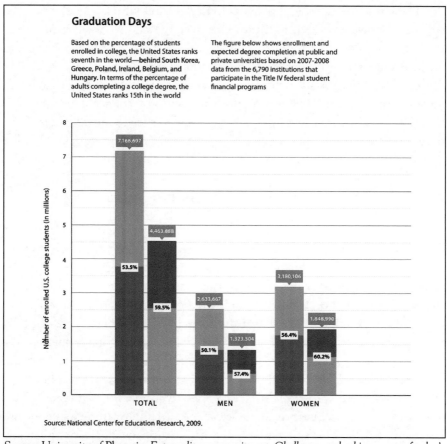

Graduation Days

Based on the percentage of students enrolled in college, the United States ranks seventh in the world—behind South Korea, Greece, Poland, Ireland, Belgium, and Hungary. In terms of the percentage of adults completing a college degree, the United States ranks 15th in the world

The figure below shows enrollment and expected degree completion at public and private universities based on 2007-2008 data from the 6,790 institutions that participate in the Title IV federal student financial programs

Source: National Center for Education Research, 2009.

Source: University of Phoenix, *Extraordinary commitment: Challenges and achievements of today's working learner*, 2010, Phoenix, AZ: Author. Retrieved from http://cdn-static.phoenix.edu/content/dam/altcloud/doc/extraordinary-commitment.pdf?cm_sp=UOPX+Knowledge+Network-_-PDF-_-Extraodinary+Commitment

While developing nations like Singapore, Korea, and China are pouring money into education, funding in the U.S. is being cut in many states. In California, for example, salaries have been cut, employees furloughed, and enrollments reduced, rather than increased to meet the coming needs of the workplace. Though the situation is most serious in California, states across the country have trimmed budgets for public education institutions (Fischer, 2009). While some authors phrase concern about this situation in terms of the status of American educational institutions (Fischer, 2009), others argue that Americans will not be able to maintain their standard of living if they do not produce enough workers to compete in the global economy (Auguste et al., 2010).

There is another consideration as well. An adult with a bachelor's degree will earn approximately one third more over the course of a lifetime than someone who does not complete college. She or he will earn about twice as much as someone with only a high school education (Brock, 2010). So, if educational opportunity is given disproportionately to the children of educated, relatively wealthy parents who can afford to pay for a college education at a traditional institution, not only will there be too few workers to meet the needs of the economy, but the social gaps between economic groups will widen and society will fail to make the best use of its human capital.

Are Educational Institutions Responding Fast Enough?

The breakneck speed of technological change in recent years has meant that higher education institutions have had to keep pace in key areas such as information exchange and student learning environments. With a few exceptions, leading universities have not addressed the change in the reality of the university, where information is now available in large quantities from any location, often for free. Critics charge that institutions of higher learning are not fully exploiting those mechanisms that make the mobile transfer of information possible and that they have been very slow to legitimize this reality to facilitate degree completion. One exception is the OpenCourseWork (OCW) initiative at the Massachusetts Institute of Technology (MIT; www.ocw.mit.edu). This university, at its own expense, put much of its curriculum online for public use (Bertrand, 2010). Other universities have made courses available for free on services like YouTube and iTunes U, but people do not receive credit for taking these courses or attending virtual lectures.

One expert argues that our modern American universities were set up to develop technology and apply it to industry and social needs, but now that technology has morphed into an instrument of change, ironically these same universities have failed to keep up with the changes that technology has brought. As a result, these institutions have begun to be irrelevant to a new global lifestyle. This lack of responsiveness to change is attributed to institutional arrogance and entrenched institutional structures (Bertrand, 2010).

Adrianna Kezar, who has studied change in higher education, has a slightly different view. She acknowledges that resistance to change, insufficient planning, and lack of vision have contributed to the problem. Nevertheless, she sees the differentiation of the modern university, with disconnected departments that communicate ineffectively with each other, as the primary problem. The university model that evolved in the environment of slow change that characterized the 19th and 20th centuries is now impeding colleges and universities that need to make rapid shifts to keep up with accelerating change outside the campus. She argues that effective reform needs some form of centralized control or coordination, so that the institutions have an overview and can act more quickly (Kezar, 2009).

Meeting the Education Challenge

The question for modern educators is whether higher education institutions in the 21st century are meeting the needs of learners and the economy. To do so is a staggering task. Change has accelerated at such a rate since the turn of the millennium that many institutions are scrambling to keep up. The consensus is that higher education has not adequately met the needs of a changing society. It is clearly not producing enough graduates to fill the new technology jobs that will open during the next decade.

In fact, rather than revving up to meet new needs, U.S. education has become stagnant. While college enrollment rates are rising in virtually every other industrialized country around the world, U.S. enrollment rates have remained at the same level for a decade—and at the same time employers have been demanding more education for available jobs (Auguste et al., 2010). This is clearly a recipe for potential economic disaster.

Institutions of higher education have so far struggled to meet the challenge of a changing society, where change is driven by the fast-paced evolution of technology and changes in family and social structures. The time for the U.S. to deal with this problem is running out if America's workforce is to produce enough trained workers over the next several decades.

In February 2009, President Barack Obama told Congress that Americans would once again have the highest percentage of college graduates in the world by 2020. To achieve that, the National Center on Higher Education Management Systems estimates that U.S. schools would have to produce an additional 8.2 million graduates by that date. At the current costs of a traditional education (which are likely to continue to increase), this would require an additional $150 billion in education spending—a burden that neither the states nor the federal government can afford in the present economic climate. Instead, a new economy in education will be required if we are to reach that goal (Carnevale et al., 2010).

Lou Anna Kimsey Simon, president of Michigan State University, is calling for universities to join a new network of what she calls *world grant universities*. This idea is inspired by the land grant program of the 19th century, the Morrill Land-Grant Acts, which helped states fund educational institutions and which brought economic development and prosperity to states that participated. In recognition of the new global society, Kimsey Simon suggests "integrating the attributes and strengths of all segments of society for the sustainable prosperity and well-being of peoples and nations throughout the world" (Bertrand, 2010, p. 114). She considers this effort a moral imperative, especially now that we have the technological means to achieve sustainable prosperity.

For the world economy to continue to function effectively, educational institutions must provide learning services for nontraditional students. Companies worldwide need to hire workers who have the technological skills and international understanding to compete in an increasingly complex global marketplace. The number of students with the ability and resources to attend college full-time right out of high school is not large enough to meet employers' future needs. With technology and information changing and expanding so quickly, a single training program will not be sufficient for a worker's career. They must continue to learn throughout their working lives to keep abreast of changes and new information about the global village in which we live.

Given these growing needs, American education is clearly at a crossroads. Several commentators have come forward to make predictions about what that education will look like in coming decades.

Daniel Yankelovich, founder of several organizations focusing on education and policy research, predicts five major trends. They include:

- A *changing attitude about life cycles*. We will no longer expect students to attend college from ages 18 to 22 and then move into the job market, he says. Students will delay education, take breaks, stretch out the process, and return for later training. The distinction between education and workplace training will fade into a continuing process of collaborative education. Yankelovich also sees baby boomers as a large market for education if the educational institutions choose to respond to their needs. These affluent potential students could be a source of funding to train new workers and address labor shortages (Yankelovich, 2005).

- A *focus on making science and technology more appealing to students*. The U.S. may not continue to be competitive in the global market, especially as competition from Asia rises, if more U.S. students are not channeled into these fields. Only 32% of U.S. students obtain degrees in science and engineering, while 66% of Japanese students and 59% of Chinese students do (Yankelovich, 2005).

- *The need for people living in a global community to understand other cultures, religious traditions, and languages.* The Cold War fueled an interest in foreign language and culture, but with the end of that era, funding for these areas has dropped. This has left the U.S. at a disadvantage in dealing, for example, with Islamic militants, who make up a tiny percentage of Muslims. Because it does not understand Islamic philosophy and religious tradition, Yankelovich believes the U.S. tends to react in ways that alienate people and that the militants can use to make America into a scapegoat. Yankelovich believes that it should be the job of colleges to train people to better understand other cultures and ways of communication, and he predicts new academic programs and curricula will arise to address those needs (Yankelovich, 2005).

- *A commitment to social mobility.* America will need all its human resources in coming decades, and it must find a way for enough people to get college educations. This access should be an equal access, not dependent on gender, race, or economic background. One possible solution Yankelovich proposes is to have employers pay some of the costs of education, on the condition that the worker spends a certain amount of time working with the company. He also predicts improved integration between high schools and colleges, so that students are better prepared for college (Yankelovich, 2005).

- *More balance between science and other ways of knowing, such as religious and philosophical ideas.* Yankelovich sees a polarization between advocates of these approaches and predicts an increase in mutual respect that will leave higher education less embattled and less at odds with advocates of moral values (Yankelovich, 2005).

Other commentators focus on the impact new technology can be expected to have on the future of higher education. The *Economist* Intelligence Unit has predicted that online learning will increase at colleges and universities around the world. It also predicts that alliances between corporations and universities

will increase and that this will change how decisions about providing education are made. It also predicts that universities will respond to globalization by
creating overseas presences and that new technology will revolutionize the way
young digital natives are educated in coming decades. This education of the
future will be individually paced and allow individual variation. Teaching will
be more focused on the student and less focused on the syllabus (The *Economist*
Intelligence Unit, 2008).

New Program Models

One possible model for the education of the future is distance learning using
electronic technology, where the entire program is taught via electronic means
such as virtual worlds, Skype, telepresence, video conferencing, virtual meetings between students and mentors, email, texting, blogs, and online videos.
This model has the advantage of being available to the student anywhere in the
world, and except for specific meetings, the educational material would also be
available at any time. This allows learners to access lectures and resource
material whenever they have a break in work and family responsibilities. They
can do this without spending any extra time commuting to classes or traveling
to find resource materials, making the e-learning distance model a good fit for
many working students and lifelong learners. This model also makes education
more accessible to a wide variety of learners, and some public institutions see
this as an essential aspect of fulfilling their missions for public service (The
Economist Intelligence Unit, 2008).

At most institutions of higher education, however, this kind of distance
learning is only an adjunct to traditional face-to-face classes. Most students surveyed in 2008 saw distance education based in e-learning as a supplement to
such traditional classes, not as a separate program. Over 60% of the students
said that degrees arising from a traditional program had more credibility than
degrees in an e-learning distance program. That perception was poised to
change, however, as top schools like Johns Hopkins and Stanford began to offer
highly regarded classes online (The *Economist* Intelligence Unit, 2008).

One MBA program is a case in point. This program is a hybrid of distance
e-learning and on-campus classes, but the distance learning segments are
entirely online. The students are all employees of one consulting firm who work
at least 45 hours per week, and they take two classes at one time. They start each
course with 16 hours of face-to-face class time on campus, spread out over five
days, and end the course with a 2-hour wrap-up on campus. In between they

receive 12 weeks of distance learning using e-learning technology that requires them to spend 7.5 hours per week on each course (Hollenbeck, Zinkhan, & French, 2005).

Some classes are hybrid, or part e-learning and part face-to-face. A hybrid class often has many learning activities online, so that the students spend less time in the classroom than in traditional classes (Porier, 2010). For example, lectures can be presented in a video format that students can listen to at any time, and they can repeat part, or all, of the lecture as needed. Files that accompany the lecture allow students to click on cited sources and access them for further research. Exercises and games can allow students to learn material and then experiment with how they apply this information. Time in the classroom, by contrast, can be reserved for discussion and questions or supervised experimentation.

As far back as 2002, Graham Spanier, the president of Pennsylvania State University, called hybrid courses "the single-greatest unrecognized trend in higher education today" (El Mansour & Mupinga, 2007, p. 2). Research conducted by the U.S. Department of Education in 2009 showed this kind of hybrid model of instruction had some of the fastest growing enrollment in higher education. One professor points out that the hybrid class model has the advantage of allowing students to organize information in the way that makes sense to them, so it is flexible and fits into many learning styles. A shy student, for example, may be more likely to text a comment to other students than to speak up in class (Porier, 2010).

Hybrid classes have the additional advantage of reducing costs by freeing up classroom space and making it possible for faculty members to address a more dispersed student body that is available at different times. Students who were asked about their experiences in hybrid classes said they liked the convenience of the time flexibility and online interactions. Some had difficulty with technical problems and felt lost in cyberspace; training for non–digital natives might solve those difficulties (El Mansour & Mupinga, 2007).

Chris Dede, a professor of learning technologies at Harvard, argued in 2002 that research indicated students actually learn better online than they do in the traditional face-to-face setting (El Mansour & Mupinga, 2007). The variety of formats available in e-learning may help to increase student interest. Class delivery can include audio, video, graphics, and animation, making the learning experience fun and stimulating. Some critics argue, however, that an absence of real-time discussion may mean that the discussions will lack depth and prevent a feeling of connection with instructors. This occurs when students and instructors post comments and reply to them (El Mansour & Mupinga. 2007). These crit-icisms, however, may not reflect the experiences of digital natives, who have grown up communicating and connecting via email, blogs, and text messages.

The Role of Business

Pundits have predicted that business will be more involved in future education as investors and sources of funding for their employees' education. As distance learning tools improve and degrees obtained through them gain in credibility and stature, we can expect to see more companies collaborating with colleges and universities to set up programs like the aforementioned distance learning MBA program for employees of the consulting firm (Hollenbeck et al., 2005).

Many corporations have already formed colleges or universities to train their employees, because higher education institutions have not been responsive to their need for better trained employees over the past four decades. There are cur-rently about 1,600 of these private universities, but few of them are accredited. Some corporations are interested in forming partnerships with traditional non-profit institutions and newer for-profit programs to enable the private universi-ties to offer degrees for the education their employees receive (Morey, 2004). This has the potential to provide much-needed revenue to education institutions and to nurture the trained employees that companies require.

As additional for-profit colleges and universities are developed, businesses may become more interested in education as a business and investment oppor-tunity, as well as a source of highly skilled workers needed for the global mar-ketplace. At the turn of the 21st century, consortia were formed by primarily British, Australian, and Canadian universities to provide educational mater-ial online to students in Third World countries. The programs mostly involved graduate and postgraduate training and were intended to generate income to support the colleges and universities in the consortia (Guri-Rosenblit, 2005).

Mainstream institutions of higher education have made forays into for-profit education investment, although most have kept these enterprises sepa-rate from the activities of their research universities. Harvard University, for example, is an investor in Charlesbank of Boston, which made a 2002 invest-ment in a for-profit training institution for automobile and motorcycle tech-nicians. Dartmouth, Johns Hopkins, and Brown have also invested in for-profit educational companies (Guri-Rosenblit, 2005).

The trend toward globalization has had a big impact on the development of for-profit education. Technology has produced a real-time, globally connected world community that is not limited by national divisions. This has led to the development of many online educational courses that may be a step toward the globalization of education. Several institutions have been identified as moving in this direction, including the University of Maryland, the British Open University, Monash University of Australia, and the University of Phoenix (Morey, 2004). This group comprises a broad spectrum of schools that are increasingly deploying e-learning technology. The University of Maryland is a high-ranking state-administered school that has large campus with extensive infrastructure, a sports program, and all the trappings of traditional higher education. The Open University was founded under a Royal Charter and receives government funds. It was set up specifically as a distance learning institution and was intended to develop programs for this type of learning. Monash University is another public university, with a campus in Melbourne, Victoria. The University of Phoenix is a subsidiary of The Apollo Group, a for-profit company that provides education for working adults and lifelong learners through several different schools, including the College for Financial Planning, the Institute for Professional Development, and Meritus University.

The inclusion of a for-profit program in this list of successful distance learning programs is indicative of the shifts taking place in higher education. For-profit programs have been growing over the past decades, and by 2004 there were 650 such schools in the U.S. This growth has been fueled by the needs of older, nontraditional students and the rising cost of traditional higher education programs. These for-profit programs have sought and received accreditation and now compete directly with traditional programs. Other successful for-profit programs include DeVry University, which provides education in business and technology and has educational sites at locations throughout the U.S. and Canada, and Jones International University, which offers certificates and bachelor's and master's degrees in various areas of business. Jones operates exclusively as a distance learning program with all of its curricula online. Enrollments are predominantly from the U.S., but they include students from 70 countries (Morey, 2004).

Implications for the Future

- **Individuals:** The results of this research have implications for individuals, educational institutions, business, and government. To be successful in the next decade, individuals will need to demon-

strate foresight in navigating a rapidly shifting landscape of organizational forms and skill requirements. They will increasingly be
called upon to continually reassess the skills they need and to
quickly gather the right resources to develop and update these
skills. Workers in the future will need to be adaptable, lifelong
learners.

- **Educational institutions** at the primary, secondary, and postsecondary levels are largely the products of the technology infrastructure and social circumstances of the past. The landscape has
  changed and educational institutions should consider how to adapt
  quickly in response. Some directions of change might include:
  - ➢ Placing additional emphasis on developing skills such as
    critical thinking, insight, and analysis capabilities;
  - ➢ Integrating new-media literacy into education programs;
  - ➢ Including experiential learning that gives prominence
    to soft skills—such as the ability to collaborate, work in
    groups, read social cues, and respond adaptively;
  - ➢ Broadening the learning constituency beyond teens and
    young adults through to adulthood; and
  - ➢ Integrating interdisciplinary training that allows stu-
    dents to develop skills and knowledge in a range of sub-
    jects.

- **Businesses** must also be alert to the changing environment, and
  must adapt their workforce planning and development strategies
  to ensure alignment with future skill requirements. Strategic
  human resource professionals might reconsider traditional methods for identifying critical skills as well as selecting and developing talent. Taking into consideration those disruptions likely to
  shape the future will allow businesses to continually reassess and
  renew the skills needed for sustainable business goals. A workforce
  strategy aimed at these business goals must be considered a priority for human resource professionals and should involve collaborations across higher education institutions with a view to
  developing highly skilled workers and lifelong learners.

- **Governmental policy makers** will need to respond to the changing landscape by taking a leadership role and making education a
  national priority. If education is not prioritized, we risk compromis-

ing our ability to prepare our people for a healthy and sustainable future. For Americans to be prepared and for our businesses to be competitive, policy makers should consider the full range of skills citizens will require, as well as the importance of lifelong learning and constant skill renewal.

Summary

The future of higher education is tied to innovation, adaptability, technological progress, and an enduring knowledge of society's transformative nature. Learning institutions must break free from the confines of traditional infrastructures because learners today are dynamic knowledge-seekers who expect their learning environments to be flexible and accessible in ways that are conducive to their social, cultural, and economic backgrounds. Developing curricula and programs that address the job skills of tomorrow is imperative, as employers today have the ability to tap into to a global talent pool. Higher education institutions must cater to the needs of lifelong learners, who will enter and reenter the education arena based on personal constraints, circumstances, and propensities. Businesses and government can aid in the effort to create a vibrant workforce through leadership, innovative thinking, and smart policies that promote lifelong learning and skills development.

References

Auguste, B. G., Cota, A., Jayaram, K., & Laboissière, M. C. A. (2010, November). Winning by degrees: The strategies of highly productive higher-education institutions. Retrieved from McKinsey & Company website: http://www.mckinsey.com/clientservice/Social_Sector/our_practices/Education/Knowledge_Highlights/~/media/Reports/SSO/Winning_by_degrees_Exec_Summary_12Nov2010b.ashx

Bertrand, W. E. (2010, Fall). Higher education and technology transfer: The effects of "technosclerosis" on development. *Journal of International Affairs, 64*(1), 101.

Brock, T. (2010, Spring). Young adults and higher education: Barriers and breakthroughs to success. *The Future of Children, 20*(1), 109–132.

Bye, D. A., Pushkar, D., & Conway, M. (2007, February). Motivation, interest, and positive affect in traditional and nontraditional undergraduate students. *Adult Education Quarterly, 57*(2), 141–158.

Carnevale, A. P., Smith, N., & Stohl, J. (2010, June). Projections of jobs and education requirements through 2018. Georgetown University Center on Education and the Workforce. Retrieved from http://www9.georgetown.edu/grad/gppi/hpi/cew/pdfs/fullreport.pdf

The *Economist* Intelligence Unit. (2008). The future of higher education: How technology will shape learning. Retrieved from http://www.nmc.org/pdf/Future-of-Higher-Ed-(NMC).pdf

El Mansour, B., & Mupinga, D. M. (2007, May). Students' positive and negative experiences in hybrid online classes. *College Student Journal, 41*(1), 242–248.

Fischer, K. (2009, October 9). As U.S. retrenches, Asia drives growth through higher education. *The Chronicle of Higher Education, 56*(7), A1–A23.

Guri-Rosenblit, S. (2005). "Distance education" and "e-learning": Not the same thing. *Higher Education, 49,* 467–493.

Hollenbeck, C. R., Zinkhan, G. M., & French, W. (2005, Summer). Distance learning trends and benchmarks: Lessons from an online MBA program. *Marketing Education Review, 15*(2), 39–52.

Kezar, A. (2009, November/December). Trends in higher education: Not enough or too much? *Change,* 18–23.

Morey, A. I. (2004). Globalization and the emergence of for-profit higher education. *Higher Education, 48,* 131–150.

Nicholas, S. (2008). *Nontraditional older students: Higher education > nontraditional older students.* EBSCO Research Starters.

Pew Hispanic Center. (2008). *Statistical portrait of the foreign-born population in the United States, 2008.* Retrieved from http://pewhispanic.org/factsheets/factsheet.php?FactsheetID=59

Porier, S. (2010, September). A hybrid course design: The best of both educational worlds. *Techniques,* 28–30.

Pusser, B., Breneman, D. W., Gansneder, B. M., Kohl, K. J., Levin, J. S., Milam, J. H., & Turner, S. E. (2007, March). Returning to learning: Adults' success in college is key to America's future. Retrieved from Lumina Foundation for Education website: http://www.luminafoundation.org /publications/ReturntolearningApril2007.pdf

University of Phoenix. (2009). Academic report: Executive summary. Retrieved from http://www.phoenix.edu/about_us/publications/academic-annual-report/2009.html

University of Phoenix. (2011, April 14). Future Work Skills 2020. Retrieved from http://www.phoenix.edu/research-institute/publications/2011/04/future-work-skills-2020.html

Yankelovich, D. (2005, November 25). Ferment and change: Higher education in 2015. *The Chronicle of Higher Education, 52*(14), B6–B9.

INDEX

ABOUT THE AUTHOR

Dr. Tracey Wilen-Daugenti is Vice President and Managing Director of the Apollo Research Institute and Visiting Scholar at Stanford University's Media X program. She has authored several works on the future of higher education, including *.edu: Technology and Learning Environments in Higher Education* and a seven-book series on women and international business.